Springer Theses

Recognizing Outstanding Ph.D. Research

For further volumes:
http://www.springer.com/series/8790

Aims and Scope

The series "Springer Theses" brings together a selection of the very best Ph.D. theses from around the world and across the physical sciences. Nominated and endorsed by two recognized specialists, each published volume has been selected for its scientific excellence and the high impact of its contents for the pertinent field of research. For greater accessibility to non-specialists, the published versions include an extended introduction, as well as a foreword by the student's supervisor explaining the special relevance of the work for the field. As a whole, the series will provide a valuable resource both for newcomers to the research fields described, and for other scientists seeking detailed background information on special questions. Finally, it provides an accredited documentation of the valuable contributions made by today's younger generation of scientists.

Theses are accepted into the series by invited nomination only and must fulfill all of the following criteria

- They must be written in good English.
- The topic should fall within the confines of Chemistry, Physics, Earth Sciences, Engineering and related interdisciplinary fields such as Materials, Nanoscience, Chemical Engineering, Complex Systems and Biophysics.
- The work reported in the thesis must represent a significant scientific advance.
- If the thesis includes previously published material, permission to reproduce this must be gained from the respective copyright holder.
- They must have been examined and passed during the 12 months prior to nomination.
- Each thesis should include a foreword by the supervisor outlining the significance of its content.
- The theses should have a clearly defined structure including an introduction accessible to scientists not expert in that particular field.

Cheng Jin

Theory of Nonlinear Propagation of High Harmonics Generated in a Gaseous Medium

Doctoral Thesis accepted by
Kansas State University, USA

Author
Dr. Cheng Jin
Department of Physics
Kansas State University
Manhattan, Kansas
USA

Supervisor
Prof. Chii-Dong Lin
Department of Physics
Kansas State University
Manhattan, Kansas
USA

ISSN 2190-5053 ISSN 2190-5061 (electronic)
ISBN 978-3-319-01624-5 ISBN 978-3-319-01625-2 (eBook)
DOI 10.1007/978-3-319-01625-2
Springer Cham Heidelberg New York Dordrecht London

Library of Congress Control Number: 2013945805

Printed on acid-free paper

Springer is part of Springer Science+Business Media (www.springer.com)

List of publications during Ph.D. thesis work[1]

[1] C. D. Lin, **Cheng Jin**, Anh-Thu Le, and R. R. Lucchese. Probing molecular frame photoelectron angular distributions via high-order harmonic generation from aligned molecules. J. Phys. B **45**, 194010 (2012).

[2] Guoli Wang, **Cheng Jin**, Anh-Thu Le, and C. D. Lin. Conditions for extracting photoionization cross sections from laser-induced high-order-harmonic spectra. Phys. Rev. A **86**, 015401 (2012).

[3] **Cheng Jin** and C. D. Lin. Comparison of high-order harmonic generation of Ar using truncated Bessel and Gaussian beams. Phys. Rev. A **85**, 033423 (2012).

[4] W. Cao, G. Laurent, **Cheng Jin**, H. Li, Z. Wang, C. D. Lin, I. Ben-Itzhak, and C. L. Cocke. Spectral splitting and quantum path study of high-harmonic generation from a semi-infinite gas cell. J. Phys. B **45**, 074013 (2012).

[5] **Cheng Jin**, Julien B. Bertrand, R. R. Lucchese, H. J. Wörner, Paul B. Corkum, D. M. Villeneuve, Anh-Thu Le, and C. D. Lin. Intensity dependence of multiple-orbital contributions and shape resonance in high-order harmonic generation of aligned N_2 molecules. Phys. Rev. A **85**, 013405 (2012).

[6] C. Trallero-Herrero, **Cheng Jin**, B. E. Schmidt, A. D. Shiner, J.-C. Kieffer, P. B. Corkum, D. M. Villeneuve, C. D. Lin, F. Légaré, and A. T. Le. Generation of broad XUV continuous high harmonic spectra and isolated attosecond pulses with intense mid-infrared lasers. J. Phys. B (fast track communication) **45**, 011001 (2012).

[7] Guoli Wang, **Cheng Jin**, Anh-Thu Le, and C. D. Lin. Influence of gas pressure on high-order-harmonic generation of Ar and Ne. Phys. Rev. A **84**, 053404 (2011).

[8] **Cheng Jin**, Anh-Thu Le, Carlos A. Trallero-Herrero, and C. D. Lin. Generation of isolated attosecond pulses in the far field by spatial filtering with an intense few-cycle mid-infrared laser. Phys. Rev. A **84**, 043411 (2011).

[9] **Cheng Jin**, Hans Jakob Wörner, V. Tosa, Anh-Thu Le, Julien B. Bertrand, R. R. Lucchese, P. B. Corkum, D. M. Villeneuve, and C. D. Lin. Separation of target structure and medium propagation effects in high-harmonic generation. J. Phys. B **44**, 095601 (2011).

[1] Most parts of this thesis have been published in [3, 5, 8, 9, 10, 12, 16, 18].

[10] **Cheng Jin**, Anh-Thu Le, and C. D. Lin. Analysis of effects of macroscopic propagation and multiple molecular orbitals on the minimum in high-order harmonic generation of aligned CO_2. Phys. Rev. A **83**, 053409 (2011).

[11] Song-Feng Zhao, **Cheng Jin**, R. R. Lucchese, Anh-Thu Le, and C. D. Lin. High-order-harmonic generation using gas-phase H_2O molecules. Phys. Rev. A **83**, 033409 (2011).

[12] **Cheng Jin**, Anh-Thu Le, and C. D. Lin. Medium propagation effects in high-order harmonic generation of Ar and N_2. Phys. Rev. A **83**, 023411 (2011).

[13] Song-Feng Zhao, Junliang Xu, **Cheng Jin**, Anh-Thu Le, and C. D. Lin. Effect of orbital symmetry on the orientation dependence of strong field tunneling ionization of nonlinear polyatomic molecules. J. Phys. B **44**, 035601 (2011).

[14] Song-Feng Zhao, **Cheng Jin**, Anh-Thu Le, and C. D. Lin. Effect of an improved molecular potential on strong-field tunneling ionization of molecules. Phys. Rev. A **82**, 035402 (2010).

[15] Song-Feng Zhao, **Cheng Jin**, Anh-Thu Le, T. F. Jiang, and C. D. Lin. Determination of structure parameters in strong-field tunneling ionization theory of molecules. Phys. Rev. A **81**, 033423 (2010).

[16] **Cheng Jin**, Anh-Thu Le, Song-Feng Zhao, R. R. Lucchese, and C. D. Lin. Theoretical study of photoelectron angular distributions in single-photon ionization of aligned N_2 and CO_2. Phys. Rev. A **81**, 033421 (2010).

[17] Song-Feng Zhao, **Cheng Jin**, Anh-Thu Le, T. F. Jiang, and C. D. Lin. Analysis of angular dependence of strong-field tunneling ionization for CO_2. Phys. Rev. A **80**, 051402 (2009).

[18] **Cheng Jin**, Anh-Thu Le, and C. D. Lin. Retrieval of target photorecombination cross sections from high-order harmonics generated in a macroscopic medium. Phys. Rev. A **79**, 053413 (2009).

[19] Van-Hoang Le, Ngoc-Ty Nguyen, **C. Jin**, Anh-Thu Le, and C. D. Lin. Retrieval of interatomic separations of molecules from laser-induced high-order harmonic spectra. J. Phys. B **41**, 085603 (2008).

Supervisor's Foreword

The interaction of light with matter is a common method for uncovering the microscopic structure of a material in the physical world. Light sources over the whole electromagnetic spectrum are used to "see" different aspects of an object. With the invention of infrared and visible lasers since the 1960s, it has long been possible to generate other lights in the nearby spectral range, using second harmonics or third harmonics, via the nonlinear interaction of lasers with suitable materials. As the laser technology advances, ultrashort high-power infrared lasers can now be focused into an even smaller volume to create electric fields that are comparable or greater than the typical electric field seen by an electron in an atom. For more than 20 years, high-order harmonics (HH) with energies in the tens to hundreds times of the energy of the infrared photon have been generated when intense lasers are focused into a gas medium. These light pulses extend from the extreme ultraviolet to soft X-rays. With suitable filters, these harmonics offer potentially useful tabletop light sources over a broad spectral region. Furthermore, the emitted harmonics are coherent. They can also be synthesized to form isolated attosecond pulses or attosecond pulse trains. Today these attosecond pulses are being used to study the dynamics of atoms, molecules and condensed media, to interrogate their evolution at the femto- and atto-second timescales.

High-order harmonics are generated by the coherent nonlinear interaction of intense infrared lasers with atoms and molecules in the gas phase. The basic principle for the generation of high harmonics from each atom or molecule is very simple. Near the peak of the laser field in each half optical cycle, an electron is ripped off from the atom or molecule. This electron is thrown into the oscillating electric field of the laser and later it may be driven back to recollide with the atomic or molecular ion left behind. If the electron, which has been accelerated by the laser field, recombines with the ion, then the excessive kinetic energy gained in the laser field is released in the form of a high-energy photon. This seemingly simple picture explains the emission of harmonics, but for more than two decades it has not been possible to carry out accurate theoretical calculations that can be compared to harmonic spectra observed experimentally. Without a quantitative theory, efforts to enhance the yields of harmonics in the laboratory have to depend on experimentation alone. This has slowed down the progress of making HH for practical applications as a useful light source.

High harmonics observed in the laboratory are generated from the coherent emission of harmonics from all the gas atoms or molecules as the intense laser propagates through the gas medium. In turn, the intense laser is modified by the gas medium as it propagates. To achieve high brightness for the harmonics, favorable phase matching conditions should be met. This is achieved by adjusting laser focusing conditions, gas jet position, gas pressure, and degrees of ionization. To account for all of these effects, it is necessary to develop a computer software package that can simulate the generated harmonic spectra. This goal was achieved by Cheng Jin, and this project forms the basis of his Ph.D. thesis.

In his thesis work, Cheng first studied harmonics generated on atomic targets. He was able to show that the harmonic spectra of Ar atoms from the theoretical simulation match very well with the measured ones. Later, he extended the simulations to molecular targets, for molecules that are randomly distributed or partially aligned in space. He was also able to show that harmonics can be generated from electrons that are ionized from the inner orbitals of the molecule when laser intensity is increased. For the first time, harmonic spectra measured in the laboratory can be accurately reproduced by theoretical simulations. To make such comparisons, interactions with experimentalists are essential since the experimental conditions are required for the simulation. His work establishes a new standard for the theoretical studies of high-order harmonic generation. It is no longer reasonable to study harmonic generation without considering propagation effects. After more than two decades, the details of harmonic generation in a gas medium can finally be simulated theoretically in view of Cheng's work.

The impact of the research carried out by Cheng Jin will be far-reaching in the coming years as experimentalists are pushing harmonic generation to higher energies, with the goal of generating soft X-rays in the water window region or even hard X-rays. For this purpose, mid-infrared long-wavelength driving lasers (2–5 microns) are being used. Because the harmonic yields drop rapidly for long-wavelength lasers, useful light sources for soft and hard X-rays can become reality and will have to depend on whether favorable phase-matching conditions with high gas pressure can be identified, or whether one can engineer synthesized waveform to enhance harmonics generated by individual atoms. In either endeavor, realistic simulations with mid-infrared lasers will be needed. The thesis work of Cheng Jin will serve as a great starting point for young researchers to enter this exciting field, to challenge the obstacles that one has to face in order to realize the goal of using harmonic generation as an all-purpose tool for new tabletop light sources.

Manhattan, Kansas, USA, June 2013 Prof. Chii-Dong Lin

Acknowledgments

First, I would like to sincerely thank my advisor, Dr. Chii-Dong Lin for his guidance, encouragement, and support all through my Ph.D. study. I am really grateful that he brings me to this exciting research field, constantly motivates me and gives me the invaluable advice. I still remember the most difficult time in my project. The propagation code was working at the low gas pressure, however, it didn't work at the high gas pressure. I could not figure out problems for a few months, and even wanted to give up. His patience and encouragement helped me going through that time. I also have learned a lot from my advisor about how to conduct the effective research and how to grab a hot topic in the field. This valuable experience will be helpful for my future life.

I would like to thank Dr. Anh-Thu Le for his continuous help since I first joined Prof. Lin's group. I have learned from him to run and write some computation programmes, which soon became an essential part in my project. He is always available for stimulated discussions, and helping me understand every concept in the field.

Next I would like to give my appreciations to other members, former and current, in Prof. Lin's group. Dr. Rui-Hua Xie, with whom I was closely working in the very beginning of my project, taught me step by step to do the programming and build up basic research tools. Dr. Turker Topcu initiated the propagation project, and nicely shared with me his experience gathered from successful and unsuccessful attempts. This was really useful and important for me to start a new project. Dr. Zhangjin Chen had been helping me in both research and life for many years. Dr. Toru Morishita gave me the permit to use his TDSE code, and helped me in numerical algorithm. Dr. Song-Feng Zhao had been my collaborator in the project of the tunneling ionization of molecules, which is an important part in my project. He kindly provided me with necessary data for my project. And we had a lot of good discussions in many interesting issues. I also have collaborated with Dr. Guoli Wang in the propagation project. I have been benefited from his efforts to investigate propagation effects under some extreme conditions, such as high intensity and high pressure. I appreciate my class fellow Junliang Xu, who is an intelligent and a careful researcher, for his help and collaboration in course studies. Dr. Wei-Chun Chu and Dr. Allison Harris discussed with me about some interesting problems. I really enjoyed these discussions.

I also want to thank Dr. Carlos A. Trallero-Herrrero for stimulating the project of the attosecond pulse generation, useful discussions, and the nice collaboration, and thank Dr. Lew Cocke and Wei Cao for sharing their ideas and stimulated discussions in the project of the harmonic spectral splitting.

Then I would like to acknowledge my collaboration with people from other institutes. I really thank Dr. Hans J. Wörner, Julien B. Bertrand, Dr. Paul B. Corkum, and Dr. David M. Villeneuve from National Research Council (NRC) in Canada for providing me experimental data and valuable communications. Without their efforts in experiments, I could not make a complete and convincing story. Dr. Valer Tosa from National Institute for R&D in Isotopic and Molecular Technologies in Romania, discussed with me a lot of technique details about the macroscopic propagation, and finally helped me setting up the propagation code. Dr. Robert R. Lucchese from Texas A&M University provided me with the photoionization code for molecular calculations and gave me an opportunity to visit him. I am grateful to them.

I also want to thank all of my Ph.D. committee members and outside chair: Dr. Uwe Thumm, Dr. Brian Washburn, Dr. Viktor Chikan, and Dr. Naiqian Zhang for serving on my committee and carefully reading my thesis. I thank the Department of Physics for taking good care of my study life and broadening my experience with various seminars and meetings. I thank all my friends in K-State for their friendship and their helps in all aspects.

Finally, I want to express my gratitude to people who are so important in my life. First, I would like to thank Prof. Xiao-Xin Zhou for stimulating my interest in research and providing me an opportunity to pursue my dream in the USA. Next I want to thank my dear, Jinping Fu, my parents and my brother for their endless support and love.

Contents

Chapter 1
Introduction to High-Order Harmonic Generation

1.1 Background

Laser[1] invented in 1960 has opened up new research areas in atomic and molecular physics [1–6]. Based on this technology, Franken et al. [7] first demonstrated the frequency doubling in a crystal in 1961, New and Ward [8] observed the third-harmonic generation in gases in 1967, and a few years later Reintjes et al. [9] generated the higher-order harmonic such as the fifth order. These harmonics were all in the perturbative region, which could be understood in the framework of n-photon excitation. The probability of absorbing n photons decreases exponentially with n, explaining the rapid decrease in the harmonic intensity with the harmonic order. Meanwhile, pulsed lasers were developed towards increasing peak powers, increasing repetition rates and decreasing pulse durations. The character of laser-atom interaction also evolved from being essentially pertubative for laser intensities below 10^{13} W/cm^2 to strongly nonperturbative for higher intensities. When the pulsed laser intensity reaches about 10^{14} W/cm^2, the electric field of laser becomes comparable to the Coulomb field seen by electrons in the proximity of their parent ion, and a dramatically nonlinear process of high-order harmonic generation (HHG) can occur, in which an intense ultrafast laser pulse at a given frequency is converted to integer multiples of this fundamental frequency in a conversion medium. The first observations of the HHG date back to the late 1980's. Indeed, the efficient photon emission in the extreme ultraviolet (XUV) range (from 10 eV up to 124 eV, corresponding to 124 nm to 10 nm, respectively), in the form of high-order harmonics was observed in Chicago (17th harmonic of a KrF laser, 1987) [10] and in Saclay (33rd harmonic of a Nd:YAG laser, 1988) [11]. The high-harmonic spectrum starts with a rapid decrease in efficiency for low-order harmonics consistent with the perturbation theory, followed by a broad plateau of nearly constant efficiency, and then an abrupt cutoff as shown in Fig. 1.1, which is quite different from the perturbative harmonics. Most of the early work concentrated on the extension of the plateau, i.e., to obtain the harmonics of higher frequency and

---

[1] Light amplification by stimulated emission of radiation.

C. Jin, *Theory of Nonlinear Propagation of High Harmonics Generated in a Gaseous Medium*, Springer Theses, DOI: 10.1007/978-3-319-01625-2_1,

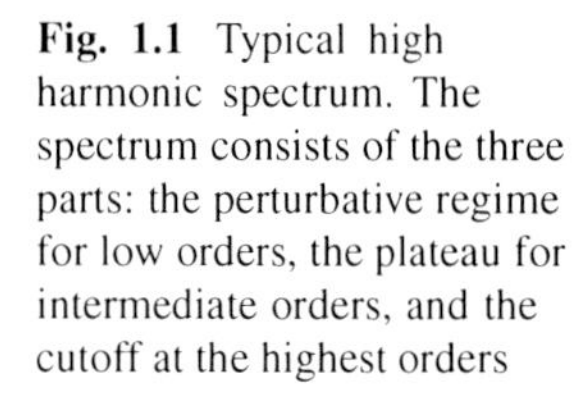

Fig. 1.1 Typical high harmonic spectrum. The spectrum consists of the three parts: the perturbative regime for low orders, the plateau for intermediate orders, and the cutoff at the highest orders

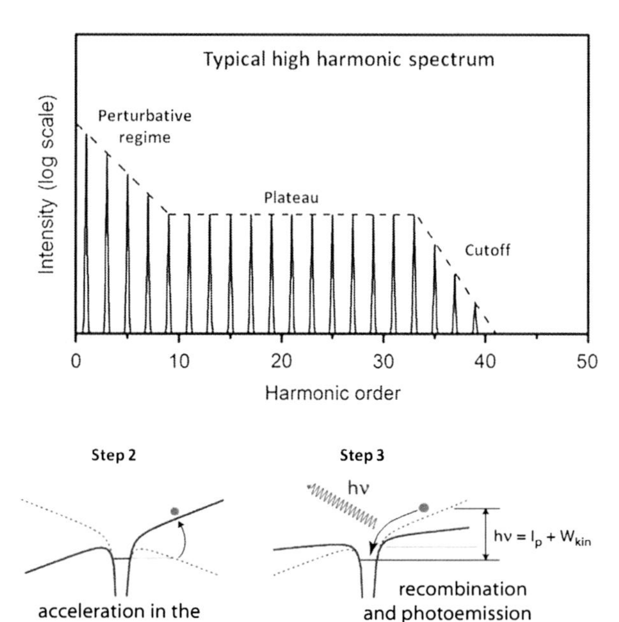

Fig. 1.2 Schematic drawing of the three-step model for HHG. Reprinted figure with permission from [21]. © (2008) by the American Physical Society

shorter wavelength. Today, HHG produced with short and intense laser pulses has been extended to the water window (below the carbon K-edge at 4.4 nm) [12–14].

Since the harmonics over a broad plateau range are with comparable efficiency, an alternative mechanism instead of the pertubative theory was required to explain HHG phenomenon. A breakthrough in the theoretical understanding was initiated in 1992–1993 by Krause et al. [15] and Corkum [16]. They have presented a semiclassical theory, which could reproduce the plateau behavior in the observed HHG spectra. According to these works, in the strong ultrafast laser field, an outmost electron is tunnel ionized through a barrier formed by the electric field and atomic potential when the electric field strength of laser is close to its peak during an optical cycle, and is driven away from the nucleus. When the oscillating laser field changes its sign (about a quarter of an optical cycle later), the electron first decelerates, then re-accelerates back towards, and finally recollides with the parent ion. Therefore, the electron can gain a significant amount of kinetic energy, much larger than the fundamental photon energy. If the returning electron recombines with its parent ion, this kinetic energy plus the ionization potential can be released in the form of the high energy photon. This model based on the "recollision picture" [17, 18] has been called as "three-step model" because there are three steps involved: ionization, propagation in the laser field and recombination, see Fig. 1.2. Krause et al. [15] also showed that the photon energy of the cutoff in the HHG spectrum followed a universal law of $I_p + 3.17U_p$, where I_p is the ionization potential, and $U_p = e^2E_0^2/4m_e\omega_0^2$ the ponderomotive energy, i.e., the mean kinetic energy acquired by an electron oscillating in the laser field. Here, e is the electron charge, m_e its mass, and E_0 and ω_0 the laser electric field and its frequency, respectively. This quasiclassical theory was shortly confirmed by a quantum-mechanical treatment also including quantum effects, such as the depletion of the ground state, the wave packet spreading and the interference based on a strong-field approximation (SFA) by Lewenstein et al. [19] and Antoine et al. [20]. Such a highly nonlinear process involved in the high harmonic generation can be accurately treated in a microscopic aspect with these simple models.

However, HHG is a process including not only the response from a single atom outlined above but also the response from a large ensemble of atoms, molecules or their ions, coherently stimulated by the laser, i.e., the response of the whole medium [1, 5, 21, 22]. Both the laser and generated harmonic fields propagate in a gaseous medium, influenced by the nonlinear effects, such as dispersion, absorption, plasma and ionization. Harmonic generation will be efficient only if the good phase-matching is achieved, requiring that the generated field to be in phase with the nonlinear polarization over the medium's length. The geometries widely used for generating HHG include that laser is focused in a gas jet or cell, and laser is guided in a hollow-core fiber or waveguide filled with gas [21, 23]. Moreover, the high laser intensity used may induce a strong ionization of the nonlinear medium. The resulting spatio- and temporal-dependent free-electron dispersion has an important consequence on the propagation of both laser and high-harmonic fields [24].

1.2 Single-Atom Response

1.2.1 Three-Step Model

The intuitive picture mentioned in Sect. 1.1 for the harmonic generation from a single atom (or the microscopic process) includes three steps: ionization, propagation and recombination. Each of them will be discussed in detail as follows:

Step 1: Ionization

In an intense laser field, the electron motion is governed by the oscillating electric field of laser pulse once it is freed (or ionized). Firstly, the electron has to escape the binding potential of an atom in the presence of an intense laser pulse. In 1965, Keldysh [25] suggested an alternative mechanism for ionization that could occur under certain conditions. At modest laser intensities ($<10^{14}$ W/cm^2), if the ionization potential is low compared with the frequency of the light (the binding energy is much bigger than the photon energy of laser) and large compared with the electric field of laser, the normal multiphoton excitation route for ionization via intermediate states applies as shown in Fig. 1.3a. When the incident field is strong enough, the atomic

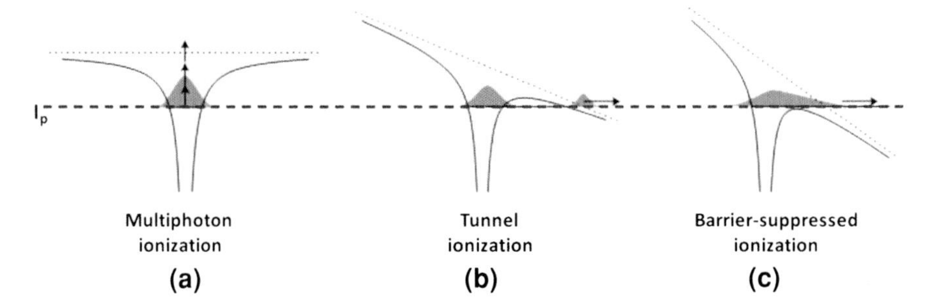

Fig. 1.3 Schematic diagram of the three ionization mechanisms. Reproduced from [5]

potential is significantly distorted to such an extent that a potential barrier is formed. If the frequency of light is low enough such that the electron can respond to this changing potential, within a quasi-stationary approximation, the electron can tunnel out through a static potential barrier as shown in Fig. 1.3b. As the laser field strength is further increased, the potential barrier is completely suppressed and the electron will classically "flow over the top" of the barrier. This is known as barrier-suppressed ionization (or over-the-barrier ionization) as shown in Fig. 1.3c. The critical field is obtained by equating the maximum induced by the field in the atomic potential to the binding energy:

$$F = \frac{\kappa^4}{16Z_c}, \tag{1.1}$$

where $\kappa = \sqrt{2I_p}$ with I_p being the binding energy, and Z_c is the charge seen by the active electron. Keldysh also introduced a parameter γ, well known as "Keldysh parameter" to determine whether the atom is ionized in the tunneling ($\gamma \ll 1$) or multiphoton regime ($\gamma \gg 1$), which was defined as:

$$\gamma = \sqrt{\frac{I_p}{2U_p}}, \tag{1.2}$$

where U_p is the ponderomotive energy as defined in Sect. 1.1. This can be understood in a qualitative way that the laser-distorted Coulomb potential oscillates with the laser frequency. For higher frequency (larger γ), the quasi-static approximation is not valid since the electron does not have enough time to accommodate the fast changes in the potential. So the motion of the electron is governed by an average over many cycles of the laser field, (i.e, the oscillating potential makes the electron bounce back and forth till it is liberated after absorbing enough photons). In the opposite limiting case (i.e., tunneling regime) (smaller γ), the tunneling time, which depends on the laser intensity and the ionization potential is larger than one optical period, so the electron has enough time to tunnel through the barrier in a single cycle. When the laser field is high enough, it can fully suppress the barrier.

Tunneling ionization model is an essential element in the theory of high harmonic generation, in which the electric field can be regarded as the quasistatic one. In 1986, Ammosov, Delone and Krainov [26] presented a generalized analytical theory, which is known as "ADK theory", to calculate the ionization rate for arbitrary atoms and initial electronic states. In 2002, Tong and Lin [27] extended the atomic ADK theory to diatomic molecules by considering the symmetry property and the asymptotic behavior of the molecular electronic wave function, and tabulated the structure parameters of several molecules for calculating ionization rates.[2] This approach is usually called molecular ADK theory, or MO-ADK theory.

Step 2: Propagation

Assume that an electron is driven away from the atomic core by a laser field $E(t) = E_0 \cos \omega_0 t$ after it is tunneled through the barrier. At later time, the laser field reverses its direction and the electron is then accelerated back towards the atomic core. Based on the Newton's law of motion, the electron displacement, x, from the core obeys [1]

$$
\begin{aligned}
\frac{d^2x}{dt^2} &= \frac{eE_0}{m} \cos \omega_0 t, \\
\frac{dx}{dt} &= v = \frac{eE_0}{m\omega_0} \sin \omega_0 t + v_i, \\
x &= -\frac{eE_0}{m\omega_0^2} \cos \omega_0 t + v_i t + x_i,
\end{aligned}
\tag{1.3}
$$

where v_i and x_i are the initial velocity and position. Assuming the conditions $v(t_0) = x(t_0) = 0$ at the moment of ionization, $t = t_0$, the different classes of electron trajectories during the propagation in the laser field are plotted in Fig. 1.4. Electrons starting from the atomic core located at (0,0), can return to the core at position 0 depending on the initial phase $\phi = \omega_0 t_0$. The final kinetic energies of electrons at the moment of recombination are determined by the intersection with the velocity axis. If the electron ionized at a phase of $\phi = 17°$ (along the cutoff trajectory b) has the highest kinetic energy of $3.17U_p$ upon it returns to the core. The electron ionized at the peak of the electric field ($\phi = 0°$) returns to the core with zero kinetic energy (can be seen in trajectory d). However, most electrons are produced at unfavorable phases of the electric field and never return to the core (for example, trajectory e). There are "short" and "long" trajectories, leading to the same returning electron energy for the harmonics in the plateau, for example, trajectories a ($\phi = 45°$) and c ($\phi = 3°$), respectively. Only first two encounters of the electron with the parent ion (as shown in Fig. 1.4) lead to significant photon emission because of the quantum-mechanical nature of the electron, which suffers from the dispersion (spreading of

[2] Their following works can be seen in [28, 29].

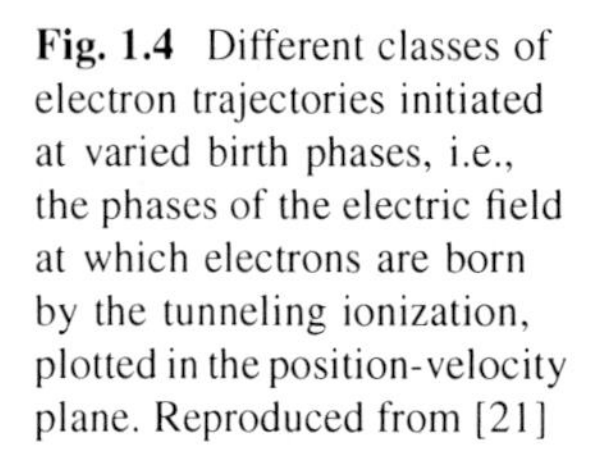

Fig. 1.4 Different classes of electron trajectories initiated at varied birth phases, i.e., the phases of the electric field at which electrons are born by the tunneling ionization, plotted in the position-velocity plane. Reproduced from [21]

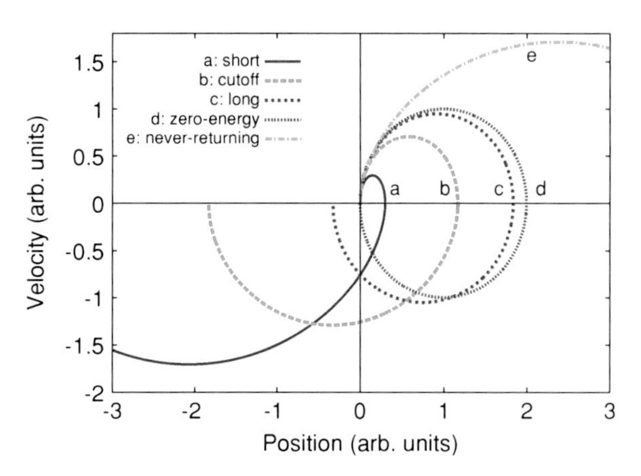

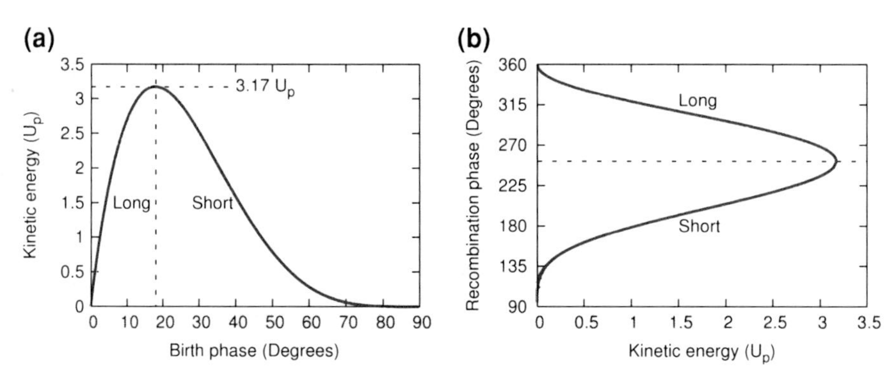

Fig. 1.5 **a** A variation of the kinetic energy of the returning electron as a function of the birth phase. **b** Plot of the recombination phase (or recombination time) changing with the final kinetic energy. The kinetic energy reaches the maximum (or cutoff) of 3.17 U_p at the birth phase of 17°. Below the cutoff, two trajectories ("short" and "long") lead to the same final kinetic energy

the wave function, thus the overlap with an atom becomes smaller) as soon as it is not bound to a potential.[3]

Step 3: Recombination

Once an electron recombines with its parent ion, the photon energy is determined by the sum of ionization potential I_p and momentary kinetic energy W_{kin} of the electron depending on the birth phase ϕ at the moment of ionization (as shown in Fig. 1.5):

$$\hbar\omega = I_p + W_{kin}, \tag{1.4}$$

[3] Tate et al. [30] claimed that there were unexpected contributions to the harmonic emission from higher-order encounters, i.e., higher-order returns, if a long-wavelength laser is applied.

where $\hbar$ is the reduced Plank constant (or Dirac constant) and ω is the angular frequency of high harmonic. The maximum kinetic energy is $3.17U_p$ as the initial phase is 17°, this determines the "cutoff" law of $I_p+3.17U_p$ [15] as mentioned in Sect. 1.1. The recombination step is replaced by two steps—the radiationless transition to the autoionizing state and the relaxation with the XUV emission, which can describe the enhancement of generation efficiency for the harmonic resonant with the transition between ground and autoionizing state of the generating ion [31]. It is also modified such that a returning electron can promote a lower-lying electron into the valance band and then recombine to the vacancy in the lower-lying state to probe the multi-electron dynamics with the high-harmonic spectroscopy [32].

High harmonic generation is only one of the strong-field nonlinear processes in the frame of three-step model. Other important processes are high-energy above-threshold ionization (HATI) [33, 34] and nonsequential double ionization (NSDI) [35, 36]. HATI originates from the elastic scattering of returning electron with the atomic ion in the backward direction. The electron gains energy in excess of its initial energy in integer multiples of the fundamental laser frequency. A typical photoelectron spectrum shows a characteristic plateau of electron peaks, separated by one fundamental photon energy, with a sharp drop around $2U_p$ and a cutoff of $10U_p$. When an inelastic collision of the electron with its parent ion occurs, another electron can be ionized so that in the end the atom is doubly ionized. This process is called "NSDI". In NSDI, a characteristic intensity dependence of doubly ionized atomic ions follows the intensity dependence of single ionization. A "knee" is observed in the intensity dependence at the point where the single ionization saturates, indicating that two processes are coupled. There are two mechanisms for removal of the second electron by the returning electron: one is through the electron-impact ionization, i.e., $(e, 2e)$ process, and the other is the electron-impact excitation followed by laser tunnel ionization [37].

1.2.2 Wavelength Scaling and Attochirp

One of the main interests in HHG studies is to produce bright tabletop XUV or soft X-ray light sources, or intense attosecond pulses. The single-atom harmonic cutoff energy is proportional to the square of laser wavelength λ_0 and intensity I_0. One could increase the cutoff energy (or extend the harmonic plateau) by increasing the laser intensity. Unfortunately, there is a practical limit at which the intensity cannot be increased due to the depletion of the ground state. An alternative way is to use longer-wavelength lasers [38, 39], while the laser intensities could be remained reasonably low. However, semiclassical strong-filed approximation [19] predicts that HHG yield from a single atom scales as λ_0^{-3}, which was partially supported by an experiment [40]. A simple physical interpretation for this scaling law is that λ_0^{-2} originates from the wave packet spreading (spending much more time in the continuum as the wavelength of laser increases) while an additional λ_0^{-1} factor arises from the conversion from photon number to energy. This is only valid for a single harmonic with both I_p/ω_0 and U_p/ω_0 constant [41]. Quantum-mechanical calculations suggested that single-atom HHG yield drops even more dramatically with the laser wavelength and scales as $\lambda_0^{-5.5}$ for the fixed laser intensity and for a same photon energy interval [30, 42, 43]. Later on, this unfavorable scaling law was partially confirmed by experimentally measured scaling laws of $\lambda_0^{-6.3\pm1.1}$ in Xe and $\lambda_0^{-6.5\pm1.1}$ in Kr at constant laser intensity (somewhat worse than the theoretical predictions) [44].[4] Actually, macroscopic dispersive effects, such as electronic, geometric, dispersion and induced dipole phase, generally result in a more rapid decrease of the HHG scaling with increasing wavelength, this will be discussed in Sect. 3.6.

HHG can also be described in an effective way based on classical trajectories. A bound electron liberated by laser field ionization can recombine with the parent ion to emit an attosecond burst of light after propagating in the laser field for approximately half an optical cycle. Since the emission process occurs twice per one optical cycle, it actually corresponds to a comb of odd-order harmonics of the fundamental driving laser field in the frequency domain. Consequently, classical mechanics as mentioned in Sect. 1.2.1 is a good approximation to describe the motion of an electron in the continuum, and the classical model predicts a dispersion of recombination times as shown in Fig. 1.5b, which corresponds to a spectral group delay dispersion (GDD) of the emitted harmonics (attochirp) [45].[5] The attochirp is the main intrinsic limitation to the duration of Fourier-synthesized attosecond pulses. In attosecond generation, the consecutive harmonics do not emit simultaneously due to the attochirp, and the ordering of frequencies defines the sign of the attochirp. Consequently, for the production of the shortest attosecond burst there exists an optimal bandwidth (Fourier transform limited pulse duration), beyond which the pulse broadens as dispersion dominates. This attochirp can be partially compensated by propagating the pulses in a suitable dispersive medium [46, 47]. In Fig. 1.6, attochirp is given by the derivative of the curve of HHG emission times as a function of harmonic photon energy. For a given class of trajectories (either "short" or "long" trajectory), the attochirp is almost a constant. It also follows that the attochirp is proportional to the ratio of the fundamental laser period to the harmonic cutoff energy [45] as shown in Fig. 1.6. Since the harmonic cutoff energy scales linearly with the laser intensity, one can reduce the attochirp by increasing the laser intensity. However, this is limited to a maximum intensity due to the depletion of the ground state as mentioned before. Another better way to reduce the attochirp consists in exploiting the wavelength scaling. Since the laser period is proportional to λ_0 and the cutoff energy scales as λ_0^2, and thus the attochirp (their ratio) should scale as λ_0^{-1}. As mentioned previously, increasing the fundamental wavelength at constant intensity can avoid the problems related to the ionization, so this method coupled with strong enhancement of the

[4] Precise physical origin of the scaling law has not been understood yet.

[5] The often used quantum mechanical treatment of the SFA [19] neglects the influence of the Coulomb potential on the motion of free electron in the laser field, and quantum paths instead of the classical trajectory involved in it with the stationary quasi-classical actions contribute mostly to the single-atom induced dipole.

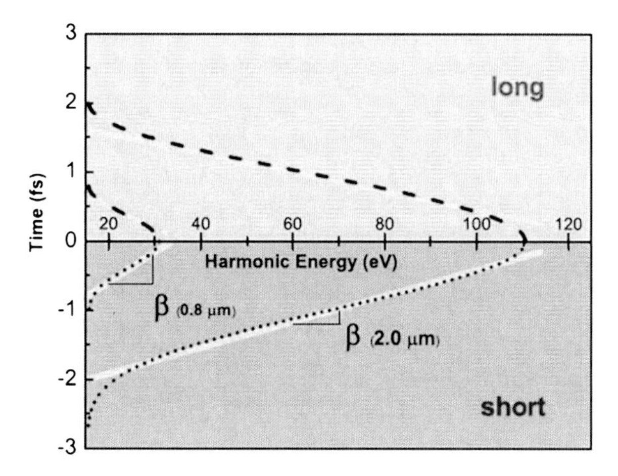

Fig. 1.6 Representation of "short" and "long" trajectories at two different wavelengths (0.8 μm and 2 μm) predicted by the semiclassical model similar to Fig. 1.5b. "Short" trajectories are emitted first, and exhibit a positive chirp β. Increasing the wavelength lowers the attochirp (indicated by the *yellow lines*, almost a constant) as evidenced by the decrease in slope. Reprinted figure with permission from [45]. © (2009) by the American Physical Society

cutoff energy would allow a better path towards shorter attosecond pulses centered at higher photon energies.

1.3 Macroscopic Propagation Effects

1.3.1 Phase Matching Conditions

The phase front of generated harmonic field should be matched with the phase front of fundamental laser field to efficiently generate the high-energy photons. In the laser-matter interaction, this phase matching could be complicated spatially and temporally due to the spatiotemporal variation of laser intensity. The relation of phase mismatch for the qth harmonic can be written as [48, 49]

$$\begin{aligned}\Delta k_q &= (k_q - qk_0) - K_{q,dip}\\ &= \Delta k_{q,geo} + \Delta k_{q,el} + \Delta k_{q,at} - K_{q,dip}.\end{aligned} \tag{1.5}$$

Here k_q and k_0 are wave vectors of harmonic and fundamental laser fields. The coherence length is proportional to the inverse of phase mismatch, and it should be larger than (or comparable to) the length of gas medium for the efficient harmonic generation. How to minimize the phase mismatch is a key issue. As shown in Eq. (1.5), there are four major sources contributing to the phase mismatch: geometric phase, electronic dispersion, atomic dispersion and induced-dipole phase. Each of them will be discussed in the following.

Geometric dispersion. To drive the HHG process, one should use an intense laser (with high enough intensity), usually obtained by confining (or focusing) a laser pulse to a small region in space, or guiding a laser beam in a waveguide. This introduces a geometric phase[6] for the fundamental laser and generated harmonics, and the phase mismatch is written as

$$\Delta k_{q,geo} = k_{q,geo}(r, z) - qk_{0,geo}(r, z). \tag{1.6}$$

There are two main geometries in practice. One is to focus a laser beam in free space. The most general mode is a Hermite-Gaussian beam. A Gaussian TEM_{00} mode is the lowest Hermite-Gaussian mode whose radial intensity distribution is Gaussian. The Gaussian beams emitted by many lasers are usually refracted by a lens to create a converging beam, and then a Gaussian beam is transformed into another Gaussian beam (characterized by a different set of parameters). However, the focal spot size of a Gaussian beam is only sustained over approximately one Rayleigh length $z_R = \pi w_0^2/\lambda_0$, where w_0 is the beam waist at the focus. After that the beam size increases approximately linearly with the propagation distance z.[7] A truncated Bessel beam is the general model for a few-cycle pulse.[8] The other geometry is to guide a laser beam in the hollow-core waveguide (or fiber) to maintain a high intensity over an extended propagation length [21]. When a laser pulse is focused into a capillary, the beam radius remains constant over the length of capillary, due to the reflection of light at the boundaries. The wave vectors of laser and the generated harmonic fields are all affected due to the particular conditions at the capillary walls. The modification of wave vector is larger if the diameter of waveguide is smaller.

Induced dipole phase. For harmonic generation in the strong-field regime, it is shown that single-atom induced dipole phase strongly depends on the laser intensity, so the spatial variation of focused laser beam results in longitudinal and transverse gradients of this phase. The contribution to the phase mismatch is

$$K_{q,dip} = \nabla \varphi_{q,dip}. \tag{1.7}$$

Here intrinsic dipole phase $\varphi_{q,dip}$ is the action accumulated by an electron during its excursion in the laser field along the trajectory leading to the emission of qth harmonic. It can be expressed as (in the first-order approximation)

$$\varphi_{q,dip} = -\alpha_i^q I, \tag{1.8}$$

where I is the instantaneous laser intensity. The proportional constant $\alpha_{i=S,\,L}$ depends on "short" (S) or "long" (L) trajectory. When the harmonics are in the plateau, $\alpha_{i=S}^q \approx 1 \times 10^{-14}$ rad cm^2/W and $\alpha_{i=L}^q \approx 24 \times 10^{-14}$ rad cm^2/W [50–53]. In the cut-off, these two trajectories merge into one, and $\alpha_{i=S,L}^q \approx 13.7 \times 10^{-14}$ rad cm^2/W. The intensity dependence of dipole phase is different for "short" and "long" trajectories, so good phase-matching conditions are consequently varied for these two trajectories. This dipole phase is also responsible for the spectral

[6] It is more generally called as "Gouy" phase.

[7] See details in Appendix D.1.

[8] See details in Appendix D.2.

broadening of high harmonics because the intensity variation $I(t)$ in time causes a frequency chirp $\Delta\omega_q(t) = -\partial\varphi_{q,dip}(t)/\partial t$.

Plasma (electronic) dispersion. The first step in the HHG process is ionization. Actually, only a small portion of electrons freed by the laser field could recombine with their parent ions to emit the high-energy photons. The rest miss the core and become free for a long time compared with the laser duration. This would finally cause the modification in the refractive index. The phase mismatch due to free electrons is

$$\begin{aligned}\Delta k_{q,el} &= k_{q,el}(r,z,t) - qk_{0,el}(r,z,t) \\ &\approx \frac{e^2 n_e(r,z,t)}{4\pi\varepsilon_0 m_e c^2} q\lambda_0 = qr_0 n_e(r,z,t)\lambda_0,\end{aligned} \tag{1.9}$$

where $n_e(r,z,t)$ is the spatiotemporal dependent electron density, e, m_e and r_0 are charge, mass and classical radius of an electron, respectively. Here the free-electron dispersion for harmonic field is neglected because the frequencies of high harmonics are much higher than the plasma frequency.

Neutral (atomic) dispersion. Any conversion media for the harmonic generation exhibit the dispersion, which is a refractive index depending on the frequency (or wavelength) of light. The phase mismatch due to neutral atom dispersion is

$$\Delta k_{q,at} = k_{q,at}(r,z,t) - qk_{0,at}(r,z,t). \tag{1.10}$$

The spatiotemporal dependence may be involved due to the ionization of medium. For fundamental laser pulse, the Sellmeier equations with coefficients that are specific for a particular medium can be applied to obtain the refractive index, where the wavelength region is not too close to a resonance. For high harmonics in the XUV (or soft X-ray) region, the refractive index is generally smaller than 1. For an intense ultrafast laser, the intensity dependent modification in the refractive index caused by the third-order susceptibility (usually called as "Kerr effect") may become important in some parts of the pulse close to laser peak.

1.3.2 Absorption Effect

For most conversion media used for harmonic generation in the XUV and soft X-ray region, the photoionization cross section σ, as well as the absorption of photons is large. So the reabsorption is a limiting factor to the phase-matched harmonics. One can define an absorption length $L_{abs} = 1/\rho\sigma$ as the length over which the intensity of XUV light propagating in an absorbing medium drops to 1/e, where ρ denotes the gas density. The harmonics are generated as the driving laser propagates along the direction z in the medium. Earlier generated harmonics are added to newly generated ones coherently, and they are also affected by the absorption. Based on a one-dimensional model, one can obtain the qth harmonic yield [54, 55]

$$I_q \propto |\int_0^{L_{med}} \rho A_q(z) \exp(-\frac{L_{med}-z}{2L_{abs}}) \exp[i\varphi_q(z)]dz|^2, \tag{1.11}$$

where $A_q(z)$ is the amplitude of microscopic single-atom response, and $\varphi_q(z)$ is its phase. For a loosely focused or guided laser beam, $A_q(z)$ can be assumed independently of z, Eq. (1.11) is written as

$$I_q \propto \rho^2 A_q^2 \frac{4L_{abs}^2}{1+4\pi^2(L_{abs}^2/L_{coh}^2)}[1+\exp(-\frac{L_{med}}{L_{abs}})-2\cos(\frac{\pi L_{med}}{L_{coh}})\exp(-\frac{L_{med}}{2L_{abs}})], \tag{1.12}$$

where $L_{coh} = \pi/\Delta k_q$ is the coherence length, calculated by using the phase mismatch Δk_q in Sect. 1.3.1. The evolution of I_q as a function of the medium length with different ratios of L_{coh}/L_{abs} is plotted in Fig. 1.7. It is easily seen from Fig. 1.7 that to generate more than half asymptotic harmonic yield for a long coherence and propagation length, the following conditions need to fulfill [54, 55]:

$$L_{med} > 3L_{abs}, L_{coh} > 5L_{abs}. \tag{1.13}$$

Under these conditions, one can even generate the harmonic yield close to the maximum, for example, harmonic yield reaches approximately 90% of the asymptotic value when L_{med} is about $6L_{abs}$. Another important feature from this analysis is that asymptotic value increases as $|A_q/\rho|^2$, independently of gas density. It requires to maximize A_q/ρ and fulfill Eq. (1.11) simultaneously to optimize the HHG yield. These optimizing conditions are also time dependent and strongly influenced by the ionization.

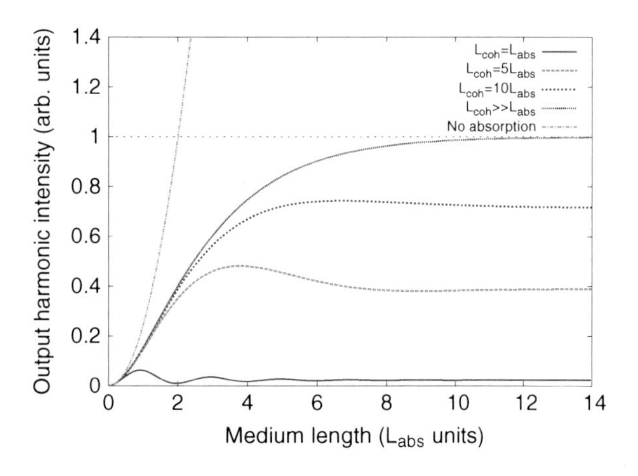

Fig. 1.7 Reabsorption of the generated HHG light in the medium. The output harmonic intensity increases quadratically with the medium length in the absence of the absorption. If the absorption is included, there is an asymptotic limit of the HHG intensity as a function of the medium length. Reproduced from [54]

1.3.3 Spatiotemporal Dynamics of Laser Pulse

Most free electrons resulted from the ionization could not recombine with their parent ions, and they create a plasma environment for laser propagation in a gaseous medium. This plasma effect contributes a negative value to the refractive index, $-e^2 n_e(r, z, t)/(2\omega_0^2 \varepsilon_0 m_e)$, with the electron density n_e [56] and other parameters defined above. The strong dependence of ionization on the laser intensity leads to spatiotemporal varied electron density, and spatiotemporal dynamics for a focused laser beam propagating in a relatively dense medium [24]. Consequently, the generated harmonics driven by the distorted laser pulse are greatly affected and shown to be good candidates for producing an isolated attosecond pulse after spatial and spectral filtering in the far field [57–59].

Defocusing. Laser focusing leads to quite different spatial distribution of the intensity (or ionization). In the beginning part of the medium, ionization probability is the highest on axis and decreases with r because of the variation of radial laser intensity, and resulting radial variation of the electron density (or refractive index) acts like a negative lens rapidly defocusing the laser beam [24]. However, the defocusing process in turn is slowed down by the resulting lower laser intensity. So most changes of the spatial profile occur in the first half of medium where the laser intensity is still high (also the high ionization). When the ionization probability becomes lower (typically a few percent), the spatial profile does not significantly change its shape and the beam diverges slowly. The laser profile close to the propagation axis is changed dramatically compared to the assumed one propagating in the vacuum (usually considered as a Gaussian beam).

Blue shift. The ultrafast laser pulse also makes laser intensity change rapidly in time. The time dependent refractive index, through the variation of electron density, leads to a time dependent phase of the electric field and therefore a frequency chirp. During the evolution of laser pulse, there are two cases for the time-dependent electron density [24]. One is that it either increases (when the intensity is high enough to induce the considerable ionization), or stays constant (when the intensity is low). The other is that intensity is sufficiently high to saturate the ionization probability, the electron density initially increases, then saturates and stays constant finally. The central frequency of laser field in both cases is blue shifted because the number of free electron is increased as a function of time, and the blue shift also has time dependence. Thus the fundamental laser field at the exit of medium is reshaped both spatially and temporally [59, 60]. The outgoing beam is much broader and more divergent than the incoming beam, and has a lower peak intensity. The peak of laser pulse may shift to off-axis position in space, it may occur earlier in time, and the effective duration of laser pulse may be shorter than incoming pulse.

1.4 Applications of High-Order Harmonic Generation

1.4.1 Generation of APT and IAP

One of the most attractive goals in nonlinear optics is to generate short-wavelength pulses with ultrashort duration [6, 61]. HHG becomes a good candidate source for this purpose due to its unique properties, such as ultrashort duration, high brightness and good coherence. The harmonic emission can be understood intuitively as ultrashort bursts emitted at each recollision of the electron with the parent ion, taking place during every half optical cycle (assuming a linearly polarized laser is applied). Each emission consists of the contributions from two electron trajectories with the shortest return times. Furthermore, either phase matching in the medium, or spatial filtering in the far field, selects the contribution of only one of these trajectories. As a result, the harmonic radiation consists of a train (called as "attosecond pulse train" – APT) of sharp short-wavelength pulses with sub-femtosecond duration, with only one burst per half cycle. Experimentally, an APT with duration of 250 as was firstly demonstrated by Paul et al. [62], which corresponded to the coherent superposition of harmonics 11–19 generated in argon. To characterize attosecond pulses, they introduced the RABITT[9] technique, in which side bands involving adjacent harmonics were measured to determine their phase relationship.

An APT is suitable for a number of specialized applications, however, for many other applications, an isolated attosecond pulse (IAP) is required to offer the unique time resolution on the attosecond timescale. Indeed, there was a plethora of techniques developed for the production of IAP, with the idea that harmonics can be limited to generate only from half an optical cycle in a few-cycle or multi-cycle infrared laser pulse. Conceptually, spectral filtering of high harmonics by a few-cycle driving pulse is the simplest scheme for producing an IAP. Generally, the harmonics in the cutoff region are emitted in one half-cycle only (will be continuous spectrum) if the driving pulse is short enough (typically two optical cycles or less) and has appropriately stabilized carrier-envelope phase (CEP). Using this method, an IAP as short as 80 attoseconds has been generated [63]. Based on the strong sensitivity of HHG process to the polarization (ellipticity) of driving laser, the harmonic emission can be localized to the time interval in which laser is linearly polarized. Using this polarization gating technique, an isolated 130 as pulse has been produced [64, 65]. The advantage of this technique is that in principle a much larger range of frequencies in the harmonic spectrum are emitted during a short time (not only the frequencies in the cutoff region), allowing for potentially much shorter attosecond pulses with higher yields. Other methods for IAP generation include spatiotemporal gating, two-color control, quasi-phase matching, and so on. In a tight-focusing geometry, different phase-matching can be achieved for different photon-energy regions. The good phase matching on axis is only for harmonic radiation within one half cycle of the driving field, and an IAP could be temporally selected by applying a spatial and

---

[9] Reconstruction of attosecond beating by interference of two-photon transitions.

spectral filter in the far field [66, 67]. The above polarization gating method is based on the use of two color fields with the orthogonal polarizations, the two-color field co-polarized is also used for IAP generation. The presence of the second field with different frequency and intensity breaks the half-cycle periodicity of the electron dynamics and offers an additional time gating. To improve the yield of harmonics, as well as the intensity of attosecond pulses, quasi-phase matching method has been used with the idea that only the constructive build-up of the harmonic radiation is allowed. The newly generated harmonics interferes with earlier ones during the propagation in the medium, while the destructive inference regions can be minimized either by modulating the generating light (for example, using counter-propagating light), or by modulating the generating material (for example, using modulated wave guides) [68–71]. Meanwhile, an IAP can be characterized using an attosecond streak camera technique pioneered by Krausz and coworkers [72, 73] relying on the use of an intense (10^{13}–10^{14} W/cm^2) infrared laser field that exchanges many photons with the electron after it has been set free in the continuum by the attosecond pulses.

The availability of an APT and IAP makes it capable of performing pump-probe experiments with the APT or IAP as a pump pulse and infrared (IR) laser as a probe pulse (usually written as "APT+IR" or "IAP+IR") [74–76]. Such experiments have an advantage that the time delay between the APT (or IAP) and the IR can be controlled with high precision at the level of attoseconds. The pump pulse can create an attosecond wave packet evolving in time, and the probe pulse can be applied at different time delays. It is of interest to observe how the results of the probe change with the time delay, and then using these results to retrieve the information on the dynamic system after the pump pulse. The availability of attosecond pulses may also allow one to perform attosecond pump-probe experiments [77] where a first XUV pulse electronically excites an atomic, molecular or condensed phase system of interest, thereby initiates an ultrafast electronic process, and a second, time-delayed XUV pulse extracts the signal from the system containing information about the time evolution that has taken place. Both pulses have a duration that is short compared to the typical timescale of the electron dynamics under investigation. Few attempts include that Hu and Collins [78] calculated two-color ionization of He using a sequence of two ultrashort XUV pulses, and Yudin et al. [79] analyzed the ionization of a set of coherently coupled states using an attoscond pulse.

1.4.2 Probing Electronic Structure and Dynamics

Harmonic generation itself has become a tool to obtain the structural and dynamical information of atoms and molecules. Because single-atom or single-molecule harmonic emission results from the interference of re-collision electron with the ground-state wave function (Step 3 as discussed in Sect. 1.2.1), the electronic structure of atomic states and molecular orbitals, or electron-electron interaction is imprinted on the HHG spectrum. In a recent experiment Shiner et al. [32] recorded high-harmonic spectra of several atoms (Kr, Xe and Ar) for the photon energy up to 160 eV using a few-cycle 1.8-μm laser. They showed that these spectra could be related to differential photoionization cross sections measured with the synchrotron sources, and these spectra contain features due to collective multi-electron effects involving inner-shell electrons, in particular the giant resonance in Xe. In addition, harmonic interferometry has been used to probe the multi-electron dynamics in molecules by Smirnova et al. [80]. They measured the phases and amplitudes of CO_2 harmonics, and revealed the features of multiple orbitals and the underlying attosecond multi-electron dynamics, including the dynamics of electron rearrangement upon ionization. In another pioneering experiment, Itatani et al. [81] demonstrated the so-called tomographic imaging of molecular orbitals, where the highest-occupied molecular-orbital (HOMO) of N_2 molecules could be retrieved by measuring the HHG spectra at a range of alignment angles of the molecular axis with respect to the polarization of driving laser. Later on, Haessler et al. [82] constructed the two highest occupied molecular orbitals (HOMO and HOMO-1) by characterizing the harmonic emission from aligned N_2 molecules. The tomographic imaging approach has also been applied to reconstruct the HOMO of CO_2 using harmonic spectroscopy [83].

Harmonic generation process can provide access to the molecular structure as well. Following the two-center interference model proposed by Lein et al. [84], Kanai et al. [85] observed the constructive and destructive interference occurring for selected harmonics subjected to a Bragg condition for the returning electron. And then one could deduce the internuclear distances in the molecule by using the harmonic order where the constructive or destructive interference occurs. Single-molecule harmonics can be viewed from an electron-ion recollision process, where the electron recollides with the positive ion predominantly during a small fraction of the optical cycle of driving laser, implying that the electron probes the parent ion with the attosecond time resolution. In an experiment on D_2^+ dissociation, one could thus map the first few femtoseconds of the molecular dissociation by using a wavelength-tunable ionization laser, where the variation of wavelength translated into a variable recollision time [86]. In another experiment, Baker et al. [87] demonstrated a method that could probe nuclear dynamics and structural rearrangement on a sub-femtosecond time scale using the HHG in molecules. The chirped nature of electron wave packet produced by the ionization in a strong field gave rise to a similar chirp in the photons emitted (with the opposite sign and a large chirp value) upon electron-ion recombination, and this chirp in the emitted light allowed one to obtain the information about nuclear dynamics with 100-as temporal resolution, from the excitation by using an 8-fs pulse, in a single laser shot.

In the above experiments the nuclear and electronic dynamics were induced by the same laser pulse, one can also apply another probe laser, which subsequently resolves the time-delayed processes of electronic and nuclear configuration of the molecule.

1.4.3 Single-Photon Ionization of Aligned Molecules

The chemical reactions and biological transformations occur on a time scale of picoseconds or less, where the time evolution of molecules can be commonly probed by exploiting the relation between the structure of a molecule and its photoabsorption spectrum [88]. The interpretation of these experiments often relies on the existing knowledge about the molecular spectroscopy. Therefore X-ray (or XUV) diffraction and electron diffraction are the conventional approaches, which serve well to image the molecular structure. In electron diffraction experiments, the wavelength of electrons is small compared to the relevant inter-nuclear distance, inducing the diffraction that enables one to resolve structures with sub-nanometer resolution. However, the creation of electron bunches shorter than 100 fs is a major challenge. Alternatively, one can use the diffraction of electrons generated within a molecule through the photoionization by an XUV or X-ray pulse.

Photoionization is the basic physical process that provides the most direct investigation of molecular structure (or electronic structure of the molecule). The ejected photoelectrons contain the information on the molecular orbitals from which they are removed. The outgoing electrons also experience the surrounding atoms in the molecule as scattering centers, endowing the photoelectron angular distribution (PAD) with the sensitivity to the underlying molecular structure. It is possible to extract the detailed information on orbitals and/or structure if the PAD is measured in the molecular frame. However, almost all earlier experimental measurements were performed from an ensemble of randomly distributed molecules. Thus the rich dynamical structures of PAD for fixed-in-space molecules still remain largely unexplored. This challenge can be met by measuring photoelectrons and fragment ions formed from the same parent molecule in coincidence. Using molecular alignment and orientation techniques, it can avoid the requirement of a rapid dissociation accompanied by an axial recoil of the fragment ions. These techniques allow the active control of angular distribution of a parent molecule before the ionization takes place using either adiabatic or nonadiabatic methods with infrared lasers [89–91].

Photoionization of aligned molecules has previously been explored with the UV/near-infrared (NIR) radiation [92–94]. However, the kinetic energy of ejected photoelectron is very low, i.e., the de Broglie wavelength of electron is much larger than the inter-atomic spacings, which is not suitable for obtaining the structural information. An alternative approach is provided by the HHG source serving as the XUV or soft X-ray light, which makes it possible to perform the single-photon ionization experiments of aligned molecules. In a few recent experiments, Thomann et al. [95] reported the angular dependence of single-photon ionization of aligned N_2 and CO_2 molecules, Kelkensberg et al. [96] measured the electron angular distribution of CO_2 molecules aligned by using a NIR laser and ionized by using the XUV pulses, and they revealed the contributions from the four highest occupied molecular orbitals and the onset of the influence of molecular structure.

1.5 Thesis Outline

In this thesis, my main focus is to develop a macroscopic propagation model to quantitatively describe the high-harmonic spectra of gaseous atoms and molecules measured in experiments. To make a complete story of the HHG, I investigate some of its important applications, i.e., probing the electronic structure of atoms and molecules, producing an isolated attosecond pulse, and photoionizing the transiently aligned molecules.

In Chap. 2, I will describe a complete theory for the HHG in a macroscopic atomic and molecular medium. The theory is divided into two parts: single-atom or single-molecule induced dipole obtained by solving the time-dependent Schrödinger equation, and macroscopic response of the medium by solving the Maxwell's wave equation for fundamental laser and high-harmonic fields.

In Chap. 3, I will apply the propagation model to simulate measured HHG spectra by using a multi-cycle laser pulse, and establish a separable approximation for the macroscopic HHG. In Chap. 4, I will extend the model to incorporate a truncated Bessel beam as an incident laser beam, which is the general case for a few-cycle laser pulse. My focus in these two chapters is Ar atom. I will especially take a close look at Cooper minimum in the HHG spectrum.

In Chap. 5, I will investigate multi-electron effect and continuum structure in the measured HHG spectrum of Xe. The reshaping of fundamental laser field plays an important role to form the continuum harmonics, and these harmonics will be proven theoretically to produce an isolated attosecond pulse.

In Chap. 6, I will simulate the HHG spectra of molecular targets by taking into account of the macroscopic propagation in the medium, which can be quantitatively compared to the measurements directly. For two examples, N_2 and CO_2 molecules, I will also discuss that multiple molecular orbitals contribute to the HHG spectrum.

In Chap. 7, I will study another important application of the HHG, i.e., single-photon ionization. Based on the well-established photoionization theory, I will calculate the photoelectron angular distribution in the laboratory frame resulting from the photoionization of aligned molecules by using the HHG source. These calculations could be compared to the future experiments.

Finally, I will summarize this thesis in Chap. 8.

References

1. M. Protopapas, C.H. Keitel, P.L. Knight, Atomic physics with super-high intensity lasers. Rep. Prog. Phys. **60**, 389–486 (1997)
2. T. Brabec, F. Krausz, Intense few-cycle laser fields: frontiers of nonlinear optics. Rev. Mod. Phys. **72**, 545–591 (2000)
3. P. Agostini, L.F. DiMauro, The physics of attosecond light pulses. Rep. Prog. Phys. **67**, 813–855 (2004)
4. A. L'Huillier, K.J. Schafer, K.C. Kulander, Theoretical aspects of intense field harmonic generation. J. Phys. B **24**, 3315–3341 (2004)

5. T. Pfeifer, C. Spielmann, G. Gerber, Femtosecond x-ray science. Rep. Prog. Phys. **69**, 443–505 (2006)
6. F. Krausz, M. Ivanov, Attosecond physics. Rev. Mod. Phys. **81**, 163–234 (2009)
7. P.A. Franken, A.E. Hill, C.W. Peters, G. Weinreich, Generation of optical harmonics. Phys. Rev. Lett. **7**, 118–119 (1961)
8. G.H.C. New, J.F. Ward, Optical third-harmonic generation in gases. Phys. Rev. Lett. **19**, 556–559 (1967)
9. J. Reintjes, R.C. Eckardt, C.Y. She, N.E. Karangelen, R.C. Elton, R.A. Andrews, Generation of coherent radiation at 53.2 nm by fifth-harmonic conversion. Phys. Rev. Lett. **37**, 1540–1543 (1976)
10. A. McPherson, G. Gibson, H. Jara, U. Johann, T.S. Luk, I.A. McIntyre, K. Boyer, C.K. Rhodes, Studies of multiphoton production of vacuum-ultraviolet radiation in the rare gases. J. Opt. Soc. Am. B **4**, 595–601 (1987)
11. M. Ferray, A. L'Huillier, X.F. Li, L.A. Lompré, G. Mainfray, C. Manus, Multiple-harmonic conversion of 1064 nm radiation in rare gases. J. Phys. B **21**, L31–L35 (1988)
12. A. Rundquist, C.G. Durfee III, Z. Chang, C. Herne, S. Backus, M.M. Murnane, H.C. Kapteyn, Phase-matched generation of coherent soft X-rays. Science **280**, 1412–1415 (1998)
13. R.A. Bartels, A. Paul, H. Green, H.C. Kapteyn, M.M. Murnane, S. Backus, I.P. Christov, Y. Liu, D. Attwood, C. Jacobsen, Generation of spatially coherent light at extreme ultraviolet wavelengths. Science **297**, 376–378 (2002)
14. T. Popmintchev, M.-C. Chen, D. Popmintchev, P. Arpin, S. Brown, S. Ališauskas, G. Andriukaitis, T. Balčiunas, O.D. Mücke, A. Pugzlys, A. Baltuška, B. Shim, S.E. Schrauth, A. Gaeta, C. Hernǎndez-García, L. Plaja, A. Becker, A. Jaron-Becker, M.M. Murnane, and H.C. Kapteyn, Bright coherent ultrahigh harmonics in the keV X-ray regime from mid-infrared femtosecond lasers. Science **336**, 1287–1291 (2012)
15. J.L. Krause, K.J. Schafer, K.C. Kulander, High-order harmonic generation from atoms and ions in the high intensity regime. Phys. Rev. Lett. **68**, 3535–3538 (1992)
16. P.B. Corkum, Plasma perspective on strong field multiphoton ionization. Phys. Rev. Lett. **71**, 1994–1991 (1993)
17. J.L. Krause, K.J. Schafer, K.C. Kulander, Calculation of photoemission from atoms subject to intense laser fields. Phys. Rev. A **45**, 4998–5010 (1992)
18. K.J. Schafer, B. Yang, L.F. DiMauro, K.C. Kulander, Above threshold ionization beyond the high harmonic cutoff. Phys. Rev. Lett. **70**, 1599–1602 (1993)
19. M. Lewenstein, Ph Balcou, M.Y. Ivanov, A. L'Huillier, P.B. Corkum, Theory of high-harmonic generation by low-frequency laser fields. Phys. Rev. A **49**, 2117–2132 (1994)
20. P. Antoine, A. L'Huillier, M. Lewenstein, P. Salières, B. Carré, Theory of high-order harmonic generation by an elliptically polarized laser field. Phys. Rev. A **53**, 1725–1745 (1996)
21. C. Winterfeldt, C. Spielmann, G. Gerber, Colloquium: optimal control of high-harmonic generation. Rev. Mod. Phys. **80**, 117–140 (2008)
22. P. Salières, A. L'Huillier, P. Antoine, M. Lewenstein, Study of the spatial and temporal coherence of high-order harmonics. Adv. At. Mol. Opt. Phys. **41**, 83–142 (1999)
23. T. Popmintchev, M.-C. Chen, P. Arpin, M.M. Murnane, H.C. Kapteyn, The attosecond nonlinear optics of bright coherent X-ray generation. Nat. Photonics **4**, 822–832 (2010)
24. M.B. Gaarde, J.L. Tate, K.J. Schafer, Macroscopic aspects of attosecond pulse generation. J. Phys. B **41**, 132001 (2008)
25. L.V. Keldysh, Ionization in the field of a strong electromagnetic wave (multiphonon absorption processes and ionization probability for atoms and solids in strong electromagnetic field). Sov. Phys. - JETP **20**, 1307–1314 (1965)
26. M.V. Ammosov, N.B. Delone, V.P. Krainov, Tunnel ionization of complex atoms and of atomic ions in an alternating electromagnetic field. Sov. Phys. - JETP **64**, 1191–1194 (1986)
27. X.M. Tong, Z.X. Zhao, C.D. Lin, Theory of molecular tunneling ionization. Phys. Rev. A **66**, 033402 (2002)
28. S.-F. Zhao, C. Jin, A.T. Le, T.F. Jiang, C.D. Lin, Determination of structure parameters in strong-field tunneling ionization theory of molecules. Phys. Rev. A **81**, 033423 (2010)

29. S.-F. Zhao, J. Xu, C. Jin, A.T. Le, C.D. Lin, Effect of orbital symmetry on the orientation dependence of strong field tunneling ionization of nonlinear polyatomic molecules. J. Phys. B **44**, 035601 (2011)
30. J. Tate, T. Auguste, H.G. Muller, P. Salières, P. Agostini, L.F. DiMauro, Scaling of wave-packet dynamics in an intense midinfrared field. Phys. Rev. Lett. **98**, 013901 (2007)
31. V. Strelkov, Role of autoionizing state in resonant high-order harmonic generation and attosecond pulse production. Phys. Rev. Lett. **104**, 123901 (2010)
32. A.D. Shiner, B.E. Schmidt, C. Trallero-Herrero, H.J. Wörner, S. Patchkovskii, P.B. Corkum, J.-C. Kieffer, F. Légaré, D.M. Villeneuve, Probing collective multi-electron dynamics in xenon with high-harmonic spectroscopy. Nature Phys. **7**, 464–467 (2011)
33. P. Agostini, F. Fabre, G. Mainfray, G. Petite, N.K. Rahman, Free-free transitions following six-photon ionization of xenon atoms. Phys. Rev. Lett. **42**, 1127–1130 (1979)
34. G.G. Paulus, W. Nicklich, H. Xu, P. Lambropoulos, H. Walther, Plateau in above threshold ionization spectra. Phys. Rev. Lett. **72**, 2851–2854 (1994)
35. D.N. Fittinghoff, P.R. Bolton, B. Chang, K.C. Kulander, Observation of nonsequential double ionization of helium with optical tunneling. Phys. Rev. Lett. **69**, 2642–2645 (1992)
36. B. Walker, B. Sheehy, L.F. DiMauro, P. Agostini, K.J. Schafer, K.C. Kulander, Precision measurement of strong field double ionization of helium. Phys. Rev. Lett. **73**, 1227–1230 (1994)
37. C.D. Lin, A.T. Le, Z. Chen, T. Morishita, R. Lucchese, Strong-field rescattering physics–self-imaging of a molecule by its own electrons. J. Phys. B **43**, 122001 (2010)
38. P. Agostini, L.F. DiMauro, Atoms in high intensity mid-infrared pulses. Contemp. Phys. **49**, 179–197 (2008)
39. C. Vozzi, M. Negro, S. Stagira, Strong-field phenomena driven by mid-infrared ultrafast sources. J. Mod. Opt. **59**, 1283–1302 (2012)
40. B. Shan, Z. Chang, Dramatic extension of the high-order harmonic cutoff by using a long-wavelength driving field. Phys. Rev. A **65**, 11804 (2001)
41. W. Becker, S. Long, J.K. McIver, Modeling harmonic generation by a zero-range potential. Phys. Rev. A **50**, 1540–1560 (1994)
42. K. Schiessla, K.L. Ishikawab, E. Persson, J. Burgdörfer, Wavelength dependence of high-harmonic generation from ultrashort pulses. J. Mod. Opt. **55**, 2617–2630 (2008)
43. M.V. Frolov, N.L. Manakov, A.F. Starace, Wavelength scaling of high-harmonic yield: threshold phenomena and bound state symmetry dependence. Phys. Rev. Lett. **100**, 173001 (2008)
44. A.D. Shiner, C. Trallero-Herrero, N. Kajumba, H.-C. Bandulet, D. Comtois, F. Légaré, M. Giguère, J.-C. Kieffer, P.B. Corkum, D.M. Villeneuve, Wavelength scaling of high harmonic generation efficiency. Phys. Rev. Lett. **103**, 073902 (2009)
45. G. Doumy, J. Wheeler, C. Roedig, R. Chirla, P. Agostini, L.F. DiMauro, Attosecond synchronization of high-order harmonics from midinfrared drivers. Phys. Rev. Lett. **102**, 093002 (2009)
46. R. López-Martens, K. Varjú, P. Johnsson, J. Mauritsson, Y. Mairesse, P. Salières, M.B. Gaarde, K.J. Schafer, A. Persson, S. Svanberg, C.-G. Wahlström, A. L'Huillier, Amplitude and phase control of attosecond light pulses. Phys. Rev. Lett. **94**, 033001 (2005)
47. Y. Mairesse, A. de Bohan, L.J. Frasinski, H. Merdji, L.C. Dinu, P. Monchicourt, P. Breger, M. Kovačev, R. Taieb, B. Carre, H.G. Muller, P. Agostini, P. Salières, Attosecond synchronization of high-harmonic soft X-rays. Science **302**, 1540–1543 (2003)
48. Ph Balcou, P. Salières, A. L'Huillier, M. Lewenstein, Generalized phase-matching conditions for high harmonics: the role of field-gradient forces. Phys. Rev. A **55**, 32043210 (1997)
49. H. Dachraoui, T. Auguste, A. Helmstedt, P. Bartz, M. Michelswirth, N. Mueller, W. Pfeiffer, P. Salières, U. Heinzmann, Interplay between absorption, dispersion and refraction in high-order harmonic generation. J. Phys. B **42**, 175402 (2009)
50. M. Lewenstein, P. Salières, A. L'Huillier, Phase of the atomic polarization in high-order harmonic generation. Phys. Rev. A **52**, 47474754 (1995)
51. Ph Balcou, A.S. Dederichs, M.B. Gaarde, A. L'Huillier, Quantum-path analysis and phase matching of high-order harmonic generation and high-order frequency mixing processes in strong laser fields. J. Phys. B **32**, 2973–2989 (1999)

52. M.B. Gaarde, F. Salin, E. Constant, Ph Balcou, K.J. Schafer, K.C. Kulander, A. L'Huillier, Spatiotemporal separation of high harmonic radiation into two quantum path components. Phys. Rev. A **59**, 13671373 (1999)
53. M.B. Gaarde, K.J. Schafer, Quantum path distributions for high-order harmonics in rare gas atoms. Phys. Rev. A **65**, 031406 (2002)
54. E. Constant, D. Garzella, P. Breger, E. Mevel, C. Dorrer, C. Le Blanc, F. Salin, P. Agostini, Optimizing high harmonic generation in absorbing gases: model and experiment. Phys. Rev. Lett. **82**, 1668–1671 (1999)
55. C.G. Durfee III, A.R. Rundquist, S. Backus, C. Herne, M.M. Murnane, H.C. Kapteyn, Phase matching of high-order harmonics in hollow waveguides. Phys. Rev. Lett. **83**, 2187–2190 (1999)
56. E. Esarey, P. Sprangle, J. Krall, A. Ting, Self-focusing and guiding of short laser pulses in ionizing gases and plasmas. IEEE J. Quantum Electron. **33**, 1879–1914 (1997)
57. M. Hentschel, R. Kienberger, Ch. Spielmann, G.A. Reider, N. Milosevic, T. Brabec, P. Corkum, U. Heinzmann, M. Drescher, F. Krausz, Attosecond metrology. Nature **414**, 509–513 (2001)
58. M.B. Gaarde, K.J. Schafer, Generating single attosecond pulses via spatial filtering. Opt. Lett. **31**, 3188–3190 (2006)
59. C. Jin, A.T. Le, C.A. Trallero-Herrero, C.D. Lin, Generation of isolated attosecond pulses in the far field by spatial filtering with an intense few-cycle mid-infrared laser. Phys. Rev. A **84**, 043411 (2011)
60. M.B. Gaarde, M. Murakami, R. Kienberger, Spatial separation of large dynamical blueshift and harmonic generation. Phys. Rev. A **74**, 053401 (2006)
61. A. Scrinzi, M. Yu Ivanov, R. Kienberger, D.M. Villeneuve, Attosecond physics. J. Phys. B **39**, R1–R37 (2006)
62. P.M. Paul, E.S. Toma, P. Breger, G. Mullot, F. Augé, Ph Balcou, H.G. Muller, P. Agostini, Observation of a train of attosecond pulses from high harmonic generation. Science **292**, 1689–1692 (2001)
63. E. Goulielmakis, M. Schultze, M. Hofstetter, V.S. Yakovlev, J. Gagnon, M. Uiberacker, A.L. Aquila, E.M. Gullikson, D.T. Attwood, R. Kienberger, F. Krausz, U. Kleineberg, Single-cycle nonlinear optics. Science **320**, 1614–1617 (2008)
64. G. Sansone, E. Benedetti, F. Calegari, C. Vozzi, L. Avaldi, R. Flammini, L. Poletto, P. Villoresi, C. Altucci, R. Velotta, S. Stagira, S. De Silvestri, M. Nisoli, Isolated single-cycle attosecond pulses. Science **314**, 443–446 (2006)
65. H. Mashiko, S. Gilbertson, C. Li, S.D. Khan, M.M. Shakya, E. Moon, Z. Chang, Double optical gating of high-order harmonic generation with carrier-envelope phase stabilized lasers. Phys. Rev. Lett. **100**, 103906 (2008)
66. C.A. Haworth, L.E. Chipperfield, J.S. Robinson, P.L. Knight, J.P. Marangos, J.W.G. Tisch, Half-cycle cutoffs in harmonic spectra and robust carrier-envelope phase retrieval. Nature Phys. **3**, 52–57 (2007)
67. T. Pfeifer, A. Jullien, M.J. Abel, P.M. Nagel, L. Gallmann, D.M. Neumark, S.R. Leone, Generating coherent broadband continuum soft-X-ray radiation by attosecond ionization gating. Opt. Express **15**, 17120–17128 (2007)
68. I.P. Christov, H.C. Kapteyn, M.M. Murnane, Quasi-phase matching of high-harmonics and attosecond pulses in modulated waveguides. Opt. Express **7**, 362–367 (2000)
69. A. Paul, R.A. Bartels, R. Tobey, H. Green, S. Weiman, I.P. Christov, M.M. Murnane, H.C. Kapteyn, S. Backus, Quasi-phase-matched generation of coherent extreme-ultraviolet light. Nature **421**, 51–54 (2003)
70. O. Cohen, X. Zhang, A.L. Lytle, T. Popmintchev, M.M. Murnane, H.C. Kapteyn, Grating-assisted phase matching in extreme nonlinear optics. Phys. Rev. Lett. **99**, 053902 (2007)
71. X. Zhang, A.L. Lytle, T. Popmintchev, X. Zhou, H.C. Kapteyn, M.M. Murnane, O. Cohen, Quasi-phase-matching and quantum-path control of high-harmonic generation using counter-propagating light. Nature Phys. **3**, 270–275 (2007)

72. J. Itatani, F. Quéré, G.L. Yudin, M. Yu. Ivanov, F. Krausz, P.B. Corkum, Attosecond streak camera. Phys. Rev. Lett. **88**, 173903 (2002)
73. M. Kitzler, N. Milosevic, A. Scrinzi, F. Krausz, T. Brabec, Quantum theory of attosecond XUV pulse measurement by laser dressed photoionization. Phys. Rev. Lett. **88**, 173904 (2002)
74. P. Johnsson, R. López-Martens, S. Kazamias, J. Mauritsson, C. Valentin, T. Remetter, K. Varjú, M.B. Gaarde, Y. Mairesse, H. Wabnitz, P. Salières, Ph Balcou, K.J. Schafer, A. L'Huillier, Attosecond electron wave packet dynamics in strong laser fields. Phys. Rev. Lett. **95**, 013001 (2005)
75. T. Remetter, P. Johnsson, J. Mauritsson, K. Varjú, Y. Ni, F. Lépine, E. Gustafsson, M. Kling, J. Khan, R. López-Martens, K.J. Schafer, M.J.J. Vrakking, A. L'Huillier, Attosecond electron wave packet interferometry. Nature Phys. **2**, 323–326 (2006)
76. P. Ranitovic, X.M. Tong, B. Gramkow, S. De, B. DePaola, K.P. Singh, W. Cao, M. Magrakvelidze, D. Ray, I. Bocharova, H. Mashiko, A. Sandhu, E. Gagnon, M.M. Murnane, H.C. Kapteyn, I. Litvinyuk, C.L. Cocke, IR-assisted ionization of helium by attosecond extreme ultraviolet radiation. New J. Phys. **12**, 013008 (2010)
77. M.F. Kling, M.J.J. Vrakking, Attosecond electron dynamics. Annu. Rev. Phys. Chem. **59**, 463–492 (2008)
78. S.X. Hu, L.A. Collins, Attosecond pump probe: exploring ultrafast electron motion inside an atom. Phys. Rev. Lett. **96**, 073004 (2006)
79. G.L. Yudin, S. Chelkowski, J. Itatani, A.D. Bandrauk, P.B. Corkum, Attosecond photoionization of coherently coupled electronic states. Phys. Rev. A **72**, 051401 (2005)
80. O. Smirnova, Y. Mairesse, S. Patchkovskii, N. Dudovich, D. Villeneuve, P. Corkum, M. Yu. Ivanov, High harmonic interferometry of multi-electron dynamics in molecules. Nature **460**, 972–977 (2009)
81. J. Itatani, J. Levesque, D. Zeidler, H. Niikura, H. Pépin, J.C. Kieffer, P.B. Corkum, D.M. Villeneuve, Tomographic imaging of molecular orbitals. Nature **432**, 867–871 (2004)
82. S. Haessler, J. Caillat, W. Boutu, C. Giovanetti-Teixeira, T. Ruchon, T. Auguste, Z. Diveki, P. Breger, A. Maquet, B. Carré, R. Taïeb, P. Salières, Attosecond imaging of molecular electronic wavepackets. Nature Phys. **6**, 200–206 (2010)
83. C. Vozzi, M. Negro, F. Calegari, G. Sansone, M. Nisoli, S. De Silvestri, S. Stagira, Generalized molecular orbital tomography. Nature Phys. **7**, 822–826 (2011)
84. M. Lein, N. Hay, R. Velotta, J.P. Marangos, P.L. Knight, Role of the intramolecular phase in high-harmonic generation. Phys. Rev. Lett. **88**, 183903 (2002)
85. T. Kanai, S. Minemoto, H. Sakai, Quantum interference during high-order harmonic generation from aligned molecules. Nature **435**, 470–474 (2005)
86. H. Niikura, F. Légaré, R. Hasbani, M. Yu. Ivanov, D.M. Villeneuve, P.B. Corkum, Probing molecular dynamics with attosecond resolution using correlated wave packet pairs. Nature **421**, 826–829 (2003)
87. S. Baker, J.S. Robinson, C.A. Haworth, H. Teng, R.A. Smith, C.C. Chirilă, M. Lein, J.W.G. Tisch, J.P. Marango, Probing proton dynamics in molecules on an attosecond time scale. Science **312**, 424–427 (2006)
88. A.H. Zewail, Femtochemistry: atomic-scale dynamics of the chemical bond using ultrafast lasers (Nobel lecture). Angew. Chem. Int. Ed. Engl. **39**, 2586–2631 (2000)
89. H. Stapelfeldt, T. Seideman, Colloquium: aligning molecules with strong laser pulses. Rev. Mod. Phys. **75**, 543–557 (2003)
90. B. Friedrich, D. Herschbach, Alignment and trapping of molecules in intense laser fields. Phys. Rev. Lett. **74**, 4623–4626 (1995)
91. F. Rosca-Pruna, M.J.J. Vrakking, Revival structures in picosecond laser-induced alignment of I_2 molecules. I. experimental results. J. Chem. Phys. **116**, 6567–6578 (2002)
92. M. Tsubouchi, B.J. Whitaker, L. Wang, H. Kohguchi, T. Suzuki, Photoelectron imaging on time-dependent molecular alignment created by a femtosecond laser pulse. Phys. Rev. Lett. **86**, 4500–4503 (2001)
93. C.Z. Bisgaard, O.J. Clarkin, G. Wu, A.M.D. Lee, O. Geßner, C.C. Hayden, A. Stolow, Time-resolved molecular frame dynamics of fixed-in-space CS_2 molecules. Science **323**, 1464–1468 (2009)

94. Y. Tang, Y.-I. Suzuki, T. Horio, T. Suzuki, Molecular frame image restoration and partial wave analysis of photoionization dynamics of NO by time-energy mapping of photoelectron angular distribution. Phys. Rev. Lett. **104**, 073002 (2010)
95. I. Thomann, R. Lock, V. Sharma, E. Gagnon, S.T. Pratt, H.C. Kapteyn, M.M. Murnane, W. Li, Direct measurement of the angular dependence of the single-photon ionization of aligned N_2 and CO_2. J. Phys. Chem. A **112**, 9382–9386 (2008)
96. F. Kelkensberg, A. Rouzée, W. Siu, G. Gademann, P. Johnsson, M. Lucchini, R.R. Lucchese, M.J.J. Vrakking, XUV ionization of aligned molecules. Phys. Rev. A **84**, 051404 (2011)

Chapter 2
Theoretical Tools

2.1 Introduction

My focus in this thesis is high-order harmonic generation (HHG) by an intense ultrafast laser field. In considering the interaction of an atom or a molecule with radiation, there are three basic processes [1, 2]: spontaneous emission in which an atom is treated as a classical oscillating charge that can make a spontaneous transition from an excited state to a state of lower energy, emitting a photon; absorption in which an atom absorbs a photon from an external radiation field, making a transition from a state of lower to a state of higher energy; and stimulated emission in which an atom can also emit a photon under the influence of an external radiation field. To theoretically describe the interaction of an atom or a molecule with an intense laser field, there are generally three approaches: classical theory, semiclassical theory and theory of quantum electrodynamics. In the classical theory, atoms or molecules can be modeled as a group of classical harmonic oscillators, and the laser field is simply treated as a classical electromagnetic field. In the semiclassical theory, the intense laser field is still treated classically by using Maxwell's wave equations, while the atomic or molecular system is described by using the quantum mechanics. In the theory of quantum electrodynamics, both intense laser field and atoms are treated quantum mechanically, and it can describe all phenomena involving charged particles interaction by means of exchange of photons.

As discussed in Chap. 1, HHG process involves the collective effect of many atoms or molecules in the medium interacting with a laser field. This means that a full description of the observed HHG spectra requires not only the treatment of the microscopic nonlinear laser-atom or -molecule interaction, but also the macroscopic propagation of harmonic radiation in the medium. Experimentally high harmonics are usually measured far away from the generating medium, harmonic radiation from the exit of the medium needs to propagate and diverge further in the vacuum. The individual time-dependent dipole induced by the strong oscillating laser field can be obtained by the semiclassical theory, and the fundamental laser field and high harmonic field are treated classically as the electromagnetic fields propagating in

C. Jin, *Theory of Nonlinear Propagation of High Harmonics Generated in a Gaseous Medium*, Springer Theses, DOI: 10.1007/978-3-319-01625-2_2,

the medium (or in the vacuum).[1] A typical HHG study thus consists of three parts: first, the calculation of microscopic single-atom or -molecule response by solving the time-dependent Schrödinger equation; second, the macroscopic propagation of fundamental and high-harmonic fields by solving the three-dimensional Maxwell's wave equations; third, the further propagation of high harmonics in the vacuum (probably involving the complicated optical system) by constructing an optical $ABCD$ matrix. Each part will be discussed in detail in the following.

2.2 Time-Dependent Schrödinger Equation

In this section, I am concerned about constructing the time-dependent Schrödinger equation (TDSE) of an atom in the laser field. Two approximate approaches, strong-field approximation (SFA) and quantitative rescattering (QRS) theory, are applied to solve it. The formulation of the TDSE for a molecular target is beyond the scope of this thesis, only two approximated solutions of TDSE based on the SFA and the QRS theory are presented. In this thesis the laser field is limited to a linearly polarized light.

2.2.1 Semiclassical Theory

The classical electromagnetic field is described by the Maxwell's wave equation. The vector potential of a monochromatic electromagnetic field (with a linear polarization) is

$$\vec{A}(\vec{r},t) = \vec{\varepsilon} A_0 \sin(\vec{k}\cdot\vec{r} - \omega_0 t), \tag{2.1}$$

where $\vec{k}$ is the wave vector. And angular frequency ω_0 and wave number k (the magnitude of $\vec{k}$) are related by

$$\omega_0 = kc, \tag{2.2}$$

where c is the speed of light. Vector potential $\vec{A}$ in the direction specified by the unit vector $\vec{\varepsilon}$ (polarization vector) has an amplitude $|A_0|$. Laser wavelength is much larger than the size of an atom, characterized by the Bohr radius, a_0, thus $k \cdot a_0 << 1$, and then a dipole approximation is applied here. Equation (2.1) can be written as

$$\vec{A}(t) = \vec{\varepsilon} A_0 \sin(\omega_0 t). \tag{2.3}$$

Using single-active electron (SAE) approximation [3, 4], i.e., all electrons in an atom are bound except the valence electron, the time-dependent Schrödinger equation of a valence electron in a laser field can be written as

[1] The influence of atoms or molecules on the external field is also included in this treatment.

$$i\hbar\frac{\partial}{\partial t}\psi(\vec{r},t) = \left[\frac{1}{2m_e}(\vec{p} - \frac{e}{c}\vec{A})^2 + V(\vec{r})\right]\psi(\vec{r},t)$$
$$= \left[\frac{1}{2m_e}(p^2 - \frac{e}{c}\vec{p}\cdot\vec{A} - \frac{e}{c}\vec{A}\cdot\vec{p} + \frac{e^2}{c^2}A^2) + V(\vec{r})\right]\psi(\vec{r},t), \quad (2.4)$$

where $V(\vec{r})$ is the atomic potential. To obtain Eq. (2.4), the Coulomb gauge has been adopted, which is defined in the following:

$$\nabla\cdot\vec{A}(t) = 0, \quad (2.5)$$

and the scalar potential has been taken as $\phi = 0$. The momentum operator $\vec{p}$ is

$$\vec{p} = -i\hbar\nabla. \quad (2.6)$$

Using the commutation relation between $\vec{p}$ and $\vec{A}$:

$$\vec{p}\cdot\vec{A} - \vec{A}\cdot\vec{p} = -i\hbar\nabla\cdot\vec{A}, \quad (2.7)$$

Eq. (2.4) can be written as

$$i\hbar\frac{\partial}{\partial t}\psi(\vec{r},t) = \left[-\frac{\hbar^2}{2m_e}\nabla^2 - \frac{e}{cm_e}\vec{A}\cdot\vec{p} + \frac{e^2}{2m_ec^2}A^2 + V(\vec{r})\right]\psi(\vec{r},t). \quad (2.8)$$

Under the dipole approximation, $\vec{A}(t)$ only depends on time, so the term with A^2 also depends on time only, which could be eliminated from Eq. (2.8) in terms of the unitary transformation:

$$\Psi_V(\vec{r},t) = \exp\left(\frac{e^2}{2i\hbar m_ec^2}\int_0^t A^2(t')dt'\right)\psi(\vec{r},t). \quad (2.9)$$

And then Eq. (2.8) can be simplified as

$$i\hbar\frac{\partial}{\partial t}\Psi_V(\vec{r},t) = \left[-\frac{\hbar^2}{2m_e}\nabla^2 - \frac{e}{cm_e}\vec{A}\cdot\vec{p} + V(\vec{r})\right]\Psi_V(\vec{r},t). \quad (2.10)$$

Equation (2.10) is the time-dependent Schrödinger equation of an atom in the velocity gauge.

Actually the TDSE in the length gauge is much preferred. One can introduce a unitary transformation

$$\Psi_V(\vec{r},t) = \exp\left(\frac{ie}{c\hbar}\vec{A}\cdot\vec{r}\right)\Psi_L(\vec{r},t), \quad (2.11)$$

Eq. (2.10) can be expressed as

$$i\hbar\frac{\partial}{\partial t}\Psi_L(\vec{r},t)=\left[-\frac{\hbar^2}{2m_e}\nabla^2+\frac{e^2}{2m_ec^2}A^2+\frac{e}{c}\frac{\partial\vec{A}}{\partial t}\cdot\vec{r}+V(\vec{r})\right]\Psi_L(\vec{r},t). \quad (2.12)$$

Using the transformation in Eq. (2.9), the term including A^2 could be eliminated. Because the effect of an electron with a magnetic field is much weaker compared to an electric field, the effect of the magnetic field can be neglected. The strength of electric field $\vec{E}(t)$ is related to the vector potential $\vec{A}(t)$ by

$$\vec{E}(t)=-\frac{1}{c}\frac{\partial\vec{A}(t)}{\partial t}. \quad (2.13)$$

Equation (2.12) can be further simplified as

$$i\hbar\frac{\partial}{\partial t}\Psi_L(\vec{r},t)=\left[-\frac{\hbar^2}{2m_e}\nabla^2+V(\vec{r})-e\vec{r}\cdot\vec{E}(t)\right]\Psi_L(\vec{r},t). \quad (2.14)$$

Equation (2.14) is the time-dependent Schrödinger equation of an atom in the length gauge. This is the equation used to study the microscopic nonlinear laser-atom interaction for harmonic radiation, and I use "$\Psi(\vec{r},t)$" without "L" to note the time-dependent wave function in the length gauge for simplicity.

Equation (2.14) can be solved by expanding the $\Psi(\vec{r},t)$ in terms of eigenfunctions, $R_{nl}(r)Y_{lm}(\hat{r})$, of the laser-free Hamiltonian, within the box of $r\in[0,r_{max}]$,

$$\Psi(\vec{r},t)=\sum_{nl}c_{nl}(t)R_{nl}(r)Y_{lm}(\hat{r}), \quad (2.15)$$

where radial functions $R_{nl}(r)$ are expanded in a discrete variable representation (DVR) basis set [5] associated with Legendre polynomials, while c_{nl} are calculated using the split-operator method [6].

2.2.2 Strong-Field Approximation

Based on strong-field approximation (SFA), Lewenstein et al. [7] proposed an analytic form to solve the time-dependent Schrödinger equation of an atom in a low-frequency laser field in 1994. Their theory recovers the semiclassical interpretation of HHG by Krause et al. [3] and Corkum [8] as discussed in Sect. 1.2.1, and also takes into account of quantum effects such as tunneling ionization, wave packet spreading (or quantum diffusion) and interferences. This approach will be called either "Lewenstein model" or "SFA" interchangeably in this thesis. SFA has also been applied to study the characteristics of HHG from molecular targets [9–16]. In the following I will derive the Lewenstein model for an atomic target in detail, and then it is extended for a molecular target straightforwardly.

There are two main assumptions in this model: (1) all bound states in an atom are neglected excepted for the ground state; (2) the electron in the continuum state is treated as a free particle without the influence of the Coulomb potential. The depletion of the ground state was neglected initially, and later on added by Antoine et al. [17].

Considering an atom in the SAE approximation under the influence of a laser field $E(t)$ with the linear polarization in the x direction, which satisfies Eq. (2.14), the time-dependent wave function can be expanded as (atomic units are used):[2]

$$|\Psi(t)\rangle = e^{iI_p t}\left[a(t)|0\rangle + \int d^3\vec{v}\, b(\vec{v},t)|\vec{v}\rangle\right], \tag{2.16}$$

where $a(t) \approx 1$ is the ground-state amplitude, $|0\rangle$ denotes the wave function of ground state, and $b(\vec{v},t)$ are the amplitudes of the corresponding continuum states. The equation for $b(\vec{v},t)$ can be written as

$$\dot{b}(\vec{v},t) = -i\left(\frac{\vec{v}^{\,2}}{2} + I_p\right)b(\vec{v},t) + E(t)\frac{\partial b(\vec{v},t)}{\partial v_x} - iE(t)d_x(\vec{v}). \tag{2.17}$$

Here $\vec{d}(\vec{v}) = \langle\vec{v}|\vec{x}|0\rangle$ is the transition dipole matrix element from the bound to the free state, and $d_x(\vec{v})$ is the component parallel to the polarization axis. Equation (2.17) can be solved exactly and $b(\vec{v},t)$ can be written in the closed form,

$$\begin{aligned} b(\vec{v},t) = -i\int_0^t & dt' E(t')d_x(\vec{v} - \vec{A}(t) + \vec{A}(t')) \\ & \times \exp\left\{-i\int_{t'}^t dt''\left[(\vec{v} - \vec{A}(t) + \vec{A}(t''))^2/2 + I_p\right]\right\}, \end{aligned} \tag{2.18}$$

where $\vec{A}(t)$ is the vector potential of laser field defined in Eq. (2.13).

To calculate the parallel component (with respect to the laser polarization) of the time-dependent dipole moment, one needs to evaluate $D(t) = \langle\Psi(t)|x|\Psi(t)\rangle$. Using Eqs. (2.16) and (2.18), one can obtain

$$D(t) = \int d^3\vec{v}\, d_x^*(\vec{v})b(\vec{v},t) + c.c.. \tag{2.19}$$

In the above formula, only the transition back to the ground state is considered, and the contribution from continuum to continuum part is neglected. The velocity operator is defined as follows[3]

---

[2] In the derivation, $e = -1$ in the atomic unit. This convention is different from what has been used by Lewenstein et al. [7], and this results in a sign change in front of $E(t)$ and $A(t)$.

[3] The velocity operator in Lewenstein et al. [7] is $\vec{v} = \vec{p} - \vec{A}(t)$. The discrepancy is due to the convention of e in the atomic unit.

$$\vec{v} = \vec{p} + \vec{A}(t), \tag{2.20}$$

the final expression of Eq. (2.19) is

$$D(t) = -i \int_0^t dt' \int d^3\vec{p} d_x^*(\vec{p}+\vec{A}(t)) \exp[-iS(\vec{p},t,t')] E(t') d_x(\vec{p}+\vec{A}(t')) + c.c., \tag{2.21}$$

where

$$S(\vec{p},t,t') = \int_{t'}^t dt'' \{[\vec{p} + \vec{A}(t'')]^2/2 + I_p\}. \tag{2.22}$$

Equation (2.21) has a clear physical interpretation, and it is actually a sum of probability amplitudes for the following processes: $E(t')d_x(\vec{p} + \vec{A}(t'))$ is the probability for an electron to make a transition to the continuum at time t' with the canonical momentum $\vec{p}$; $\exp[-iS(\vec{p},t,t')]$ is a phase factor when the electronic wave function is propagated from time t' until t, where $S(\vec{p},t,t')$ is the quasi-classical action; $d_x^*(\vec{p} + \vec{A}(t))$ is a transition amplitude when the electron recombines to the parent ion at time t.

The major contribution to the integral over $\vec{p}$ in Eq. (2.21) is from the stationary points of the classical action,

$$\nabla_{\vec{p}} S(\vec{p},t,t') = 0, \tag{2.23}$$

so the integral over $\vec{p}$ could be performed using a saddle-point method. Defining the electron return (or excursion) time $\tau = t - t'$, Eq. (2.21) is written as

$$\begin{aligned} D(t) = &-i \int_0^\infty d\tau \left[\frac{\pi}{\epsilon + i\tau/2}\right]^{3/2} d_x^*(\vec{p}_{st} + \vec{A}(t)) \exp[-iS(\vec{p}_{st},t,\tau)] \\ &\times E(t-\tau) d_x(\vec{p}_{st} + \vec{A}(t-\tau)) + c.c., \end{aligned} \tag{2.24}$$

where

$$\vec{p}_{st} = -\int_{t-\tau}^t dt'' \vec{A}(t'')/\tau, \tag{2.25}$$

$$S(\vec{p}_{st},t,\tau) = I_p\tau - \frac{1}{2}\vec{p}_{st}^{\,2}\tau + \frac{1}{2}\int_{t-\tau}^t \vec{A}^2(t'')dt''. \tag{2.26}$$

Here the small positive constant of ϵ comes from the regularized Gaussian integration over $\vec{p}$ around the saddle-point.

It is straightforward to put the depletion of ground state in Eq. (2.24), it becomes

$$\begin{aligned} D(t) = &-i \int_0^\infty d\tau \left[\frac{\pi}{\epsilon + i\tau/2}\right]^{3/2} d_x^*(\vec{p}_{st} + \vec{A}(t)) a^*(t) \exp[-iS(\vec{p}_{st},t,\tau)] \\ &\times E(t-\tau) d_x(\vec{p}_{st} + \vec{A}(t-\tau)) a(t-\tau) + c.c.. \end{aligned} \tag{2.27}$$

The ground-state amplitude $a(t)$ can be expressed as [9, 17]

$$a(t) = \exp\left(-\int_0^t \frac{\gamma(t')}{2} dt'\right). \tag{2.28}$$

Here $\gamma(t')$ is the ionization rate.

Equation (2.27) is the generally used formula of Lewenstein model to calculate the single-atom harmonic radiation induced by an intense laser field. For single-molecule HHG, I assume that linear molecules are aligned along the x axis, while laser field $E(t)$ is linearly polarized on the x-y plane with an angle θ with respect to the molecular axis. The parallel component of time-dependent dipole moment can be expressed as [9, 10]

$$\begin{aligned} D_{\parallel}(t) = & -i \int_0^{\infty} d\tau \left[\frac{\pi}{\epsilon + i\tau/2}\right]^{3/2} [\cos\theta d_x^*(t) + \sin\theta d_y^*(t)] a^*(t) \\ & \times [\cos\theta d_x(t-\tau) + \sin\theta d_y(t-\tau)] a(t-\tau) E(t-\tau) \\ & \times \exp[-iS(\vec{p}_{st}, t, \tau) + c.c., \end{aligned} \tag{2.29}$$

where $\vec{d}(t) = \vec{d}\,[\vec{p}_{st}(t,\tau) + \vec{A}(t)]$ and $\vec{d}(t-\tau) = \vec{d}\,[\vec{p}_{st}(t,\tau) + \vec{A}(t-\tau)]$ are transition dipole matrix elements. The perpendicular component $D_{\perp}(t)$ can be given in a similar formula with $[\cos\theta d_x^*(t) + \sin\theta d_y^*(t)]$ replaced by $[\sin\theta d_x^*(t) - \cos\theta d_y^*(t)]$ in Eq. (2.29).

Equation (2.29) is the extended formula of Lewenstein model to calculate the single-molecule HHG induced by an intense laser field. In the SFA, the transition dipole moment $\vec{d}(\vec{p})$ is given by $\langle \vec{p}|\vec{r}|0\rangle$ with the continuum state approximated by a plane wave $|\vec{p}\rangle$. For hydrogenlike atoms, the dipole matrix element for transition from the ground state to a continuum state is given in the form

$$d(p) = i\frac{2^{7/2}(2I_p)^{5/4}}{\pi} \frac{p}{(p^2 + 2I_p)^3}. \tag{2.30}$$

For other atoms and molecules, $\vec{d}(\vec{p})$ is calculated numerically with the known wave function of ground state.[4]

2.2.3 *Quantitative Rescattering Theory*

Lewenstein model is well-known to give good results for the harmonics with high-photon energies, especially for the cutoff harmonics, but it fails to predict harmonics in the lower plateau since the effect of the Coulomb potential is not included. To

[4] In the calculation, the ground-state electronic wave function of an atom or a molecule can be obtained by using quantum chemistry codes such as GAMESS or GAUSSIAN [18].

improve the SFA, a quantitative rescattering (QRS) theory has been developed in Prof. Lin's group [19–21]. The QRS theory states that single-atom or -molecule HHG can be expressed as a product of a returning electron wave packet and a photorecombination cross section (PRCS) of laser-free continuum electron back to the initial bound state. Based on the Lewenstein model, each step in the three-step model can be quantified, see Eq. (2.21). The last step, i.e., recombination, is not described precisely due to a plane-wave approximation, but the motion of an electron after tunneling ionization governed mostly by the laser field has been well treated in the SFA. One can extract an accurate returning electron wave packet by using the SFA. The PRCS as the other integral part in the QRS theory is only determined by the structure of an atomic or molecular target, and it can be accurately calculated by solving the stationary Schrödinger equation. Equations (2.27) and (2.29) for calculating the HHG are expressed in time domain, while the QRS theory deals with the induced dipole moment in frequency domain. A detailed discussion of the QRS theory for HHG is given in [20]. In the following I will briefly describe the QRS theory for atomic and molecular targets separately.

1. Atomic Target

According to the QRS theory, the induced dipole moment $D(\omega)$ of an atomic target can be written as [22]

$$D(\omega) = W(\omega)d(\omega), \tag{2.31}$$

where $d(\omega)$ is the complex photorecombination (PR) transition dipole matrix element, and $W(\omega)$ is the complex microscopic electron wave packet. $|W(\omega)|^2$ describes the flux of returning electrons and it is the property of laser only. The QRS replaces the plane wave used in the SFA by an accurate scattering wave in the calculation of PR transition dipole matrix elements, while the microscopic returning electron wave packet is the same as that in the SFA. The harmonic frequency ω is related to the electron momentum p by

$$\hbar\omega = \frac{p^2}{2m_e} + I_p. \tag{2.32}$$

In practical applications, the QRS obtains the induced dipole moment by

$$D^{\text{QRS}}(\omega) = D^{\text{SFA}}(\omega)\frac{d^{\text{QRS}}(\omega)}{d^{\text{SFA}}(\omega)}, \tag{2.33}$$

where both $D^{\text{SFA}}(\omega)$ and $d^{\text{QRS}}(\omega)$ are complex numbers, while $d^{\text{SFA}}(\omega)$ is either a pure real or pure imaginary number. Within the SAE approximation, $d^{\text{QRS}}(\omega)$ is calculated by using "exact" numerical wave functions for bound and continuum states. For Ar, the model potential is given by Müller [23],

$$V(r) = -[1 + Ae^{-Br} + (17 - A)e^{-Cr}]/r, \tag{2.34}$$

with $A = 5.4$, $B = 1$ and $C = 3.682$. In this model, the spin-orbit interaction is neglected. Parameters have been chosen such that the minimum in the photoionization (or photorecombination) cross section is reproduced correctly. Tong and Lin [24] proposed a model potential for rare atoms,

$$V(r) = -\frac{Z_c + a_1 e^{-a_2 r} + a_3 r e^{-a_4 r} + a_5 e^{-a_6 r}}{r}, \tag{2.35}$$

where Z_c is the charge seen by the active electron asymptotically, and $a_1, ..., a_6$ are parameters obtained by fitting $V(r)$ to the numerical potential from the self-interaction free density functional theory. Note that in principle the parameters in Eq. (2.33) can be generalized to many-electron wave functions if needed.

2. Molecular Target

Within the QRS theory, the induced dipole moment $D(\omega, \theta)$ of a fixed-in-space molecule is given explicitly by[5]

$$D(\omega, \theta) = N(\theta)^{1/2} W(\omega) d(\omega, \theta), \tag{2.36}$$

where $N(\theta)$ is the alignment-dependent ionization probability, $W(\omega)$ is the microscopic electron wave packet, and $d(\omega, \theta)$ is the alignment-dependent transition dipole (complex in general). Here θ is the angle between the molecular axis with respect to the laser's polarization. Only the linear molecules are investigated, and the parallel component of HHG with respect to the laser polarization is considered in this thesis. Thus only the parallel component of transition dipole $d(\omega, \theta)$ is needed in the calculation. Note that $W(\omega)$ does not depend on the alignment angle θ.

The wave packet $W(\omega)$ can be calculated in two ways. First, it can be calculated formally as

$$W(\omega) = \frac{D(\omega, \theta)}{N(\theta)^{1/2} d(\omega, \theta)}. \tag{2.37}$$

Here $D(\omega, \theta)$ for a fixed alignment angle θ can be calculated by using Eq. (2.29). And then $N(\theta)$ and $d(\omega, \theta)$ are also calculated in the frame of the SFA, where the continuum waves are replaced by the plane waves. Since the wave packet $W(\omega)$ is independent of the alignment angle θ, it could be calculated only once for a given angle θ. The second approach of obtaining the wave packet is to use a reference atom with a similar ionization potential. For the reference atom, one can perform either TDSE or SFA to calculate the induced dipole moment $D(\omega)$, and then following Eq. (2.31) the wave packet can be expressed as

$$W(\omega) = \frac{D^{ref}(\omega)}{d^{ref}(\omega)}. \tag{2.38}$$

[5] The induced dipole moment is actually expressed in the molecular (or body-fixed) frame.

Note that the ionization probability of an atom is incorporated into $W(\omega)$. In the QRS theory, single-molecule induced dipole moment is then obtained from Eq. (2.36) by combining the electron wave packet with the accurate $d(\omega, \theta)$ obtained from the quantum chemistry code [25, 26] and the tunneling ionization rate $N(\theta)$ obtained from the MO-ADK theory [27].

It has been well documented that QRS results are nearly as accurate as those obtained from the TDSE whenever the accurate results from the latter can be obtained, i.e., including the atom in the SAE approximation [22] and the H_2^+ molecular ion [28]. Applications of the QRS for HHG from single molecules have been investigated in [20, 29, 30].

2.3 Maxwell's Wave Equation

2.3.1 Fundamental Laser Field in an Atomic Target

In a dense and ionizing gaseous medium, the macroscopic propagation of a driving laser pulse is affected by refraction, nonlinear self-focusing, ionization and plasma defocusing.[6] The pulse evolution in such an atomic medium is usually described by a three-dimensional (3-D) Maxwell's wave equation [31–33]:

$$\nabla^2 E_1(r, z, t) - \frac{1}{c^2}\frac{\partial^2 E_1(r, z, t)}{\partial t^2} = \mu_0 \frac{\partial J_{\text{abs}}(r, z, t)}{\partial t} + \frac{\omega_0^2}{c^2}(1 - \eta_{\text{eff}}^2) E_1(r, z, t), \tag{2.39}$$

where $E_1(r, z, t)$ is the transverse electric field of fundamental laser pulse with the angular frequency ω_0. In cylindrical coordinates, $\nabla^2 = \nabla_\perp^2 + \partial^2/\partial z^2$, where z is the axial propagation direction. The effective refractive index η_{eff} of gas medium is written as

$$\eta_{\text{eff}}(r, z, t) = \eta_0(r, z, t) + \eta_2 I(r, z, t) - \frac{\omega_p^2(r, z, t)}{2\omega_0^2}. \tag{2.40}$$

The first term $\eta_0 = 1 + \delta_1 - i\beta_1$ takes refraction (δ_1) and absorption (β_1) effects of neutral atoms into account, the second term accounts for the optical Kerr nonlinearity depending on the laser intensity $I(t)$, and the third term is from the free electrons, which contains the plasma frequency $\omega_p = [e^2 n_e(t)/(\varepsilon_0 m_e)]^{1/2}$, where m_e and e are mass and charge of an electron, respectively, and $n_e(t)$ is the free electron density. The absorption term $J_{\text{abs}}(t)$ due to ionization is expressed as [34, 35]

$$J_{\text{abs}}(t) = \frac{\gamma(t) n_e(t) I_p E_1(t)}{|E_1(t)|^2}, \tag{2.41}$$

[6] The propagation of laser pulse in an isotropic molecular medium is similar to that in an atomic medium. In general, the optical properties are anisotropic, so the laser evolution in a partially aligned molecular medium should also take the alignment effects into account.

where $\gamma(t)$ is the ionization rate, and I_p is the ionization potential. This term is usually small under the conditions for high-harmonic generation [34, 35].

The absorption effect (β_1) on the fundamental laser pulse caused by neutral atoms is in general small, so it is neglected. Only real terms are kept in the refractive index η_{eff}, and Eq. (2.39) can be written as

$$\nabla^2 E_1(r,z,t) - \frac{1}{c^2}\frac{\partial^2 E_1(r,z,t)}{\partial t^2} = \mu_0 \frac{\partial J_{\text{abs}}(r,z,t)}{\partial t} + \frac{\omega_{\text{p}}^2}{c^2} E_1(r,z,t) - 2\frac{\omega_0^2}{c^2}(\delta_1 + \eta_2 I) E_1(r,z,t). \tag{2.42}$$

By going to a moving coordinate frame ($z' = z$ and $t' = t - z/c$) and neglecting $\partial^2 E_1/\partial z'^2$ since the z' dependence of electric field is very slow, i.e., the slowly varying envelope approximation is applied, one can obtain [36]

$$\nabla_\perp^2 E_1(r,z',t') - \frac{2}{c}\frac{\partial^2 E_1(r,z',t')}{\partial z' \partial t'} = \mu_0 \frac{\partial J_{\text{abs}}(r,z',t')}{\partial t'} + \frac{\omega_{\text{p}}^2}{c^2} E_1(r,z',t') - 2\frac{\omega_0^2}{c^2}(\delta_1 + \eta_2 I) E_1(r,z',t'). \tag{2.43}$$

The temporal derivative in Eq. (2.43) can be eliminated by a Fourier transform, yielding the equation

$$\nabla_\perp^2 \tilde{E}_1(r,z',\omega) - \frac{2i\omega}{c}\frac{\partial \tilde{E}_1(r,z',\omega)}{\partial z'} = \tilde{G}(r,z',\omega), \tag{2.44}$$

where

$$\tilde{E}_1(r,z',\omega) = \hat{F}[E_1(r,z',t')], \tag{2.45}$$

and

$$\tilde{G}(r,z',\omega) = \hat{F}\{\mu_0 \frac{\partial J_{\text{abs}}(r,z',t')}{\partial t'} + \frac{\omega_{\text{p}}^2}{c^2} E_1(r,z',t') - 2\frac{\omega_0^2}{c^2}[\delta_1 + \eta_2 I(r,z',t')] E_1(r,z',t')\}, \tag{2.46}$$

where $\hat{F}$ is the Fourier transform operator acting on temporal coordinate.

The plasma frequency $\omega_{\text{p}}(r,z',t')$ is determined by free-electron density $n_{\text{e}}(t')$, which can be calculated as following

$$n_{\text{e}}(r,z',t') = n_0\{1 - \exp[-\int_0^{t'} \gamma(r,z',\tau) d\tau]\}, \tag{2.47}$$

where n_0 is the neutral atom density, and $\gamma(r, z', \tau)$ is the ionization rate calculated from Ammosov-Delone-Krainov (ADK) theory [24, 27, 37]. The refraction coefficient δ_1, depending on the pressure and the temperature of gas medium, is obtained from the Sellmeier equation [38, 39]. The nonlinear refractive index η_2, also depending on the pressure of gas medium, can be calculated through the third-order susceptibility $\chi^{(3)}$, which can be measured experimentally [40, 41]. Note that the relationship between η_2 and $\chi^{(3)}$ in Koga et al. [42] differs from that in Boyd [43] since the latter is derived by using the time-averaged intensity of optical field.

At the entrance of a gas jet ($z' = z_{\text{in}}$), the fundamental laser field is assumed to be either Gaussian or truncated Bessel in space, and in time it has a Gaussian or cosine-squared envelop.[7] These will be specified whenever the calculated results are present. The pressure is assumed constant within the gas jet in this thesis. It can be also assumed that the atomic density distribution follows a Lorentzian [44] or Gaussian [45] profile along the propagation direction.

2.3.2 High-Harmonic Field of an Atomic Target

The 3-D propagation equation of high-harmonic field in an atomic medium is described by [36, 34, 46]

$$\nabla^2 E_{\text{h}}(r, z, t) - \frac{1}{c^2}\frac{\partial^2 E_{\text{h}}(r, z, t)}{\partial t^2} = \mu_0 \frac{\partial^2 P(r, z, t)}{\partial t^2}, \tag{2.48}$$

where $P(r, z, t)$ is the polarization depending upon the applied optical field $E_1(r, z, t)$. In this equation, the free-electron dispersion is neglected because the frequencies of high harmonics are much higher than the plasma frequency. Again going to a moving coordinate frame and neglecting $\partial^2 E_{\text{h}}/\partial z'^2$ (applying the slowly varying envelope approximation), Eq. (2.48) becomes

$$\nabla_\perp^2 E_{\text{h}}(r, z', t') - \frac{2}{c}\frac{\partial^2 E_{\text{h}}(r, z', t')}{\partial z' \partial t'} = \mu_0 \frac{\partial^2 P(r, z', t')}{\partial t'^2}. \tag{2.49}$$

One can eliminate the temporal derivative by a Fourier transform, obtaining the equation

$$\nabla_\perp^2 \tilde{E}_{\text{h}}(r, z', \omega) - \frac{2i\omega}{c}\frac{\partial \tilde{E}_{\text{h}}(r, z', \omega)}{\partial z'} = -\omega^2 \mu_0 \tilde{P}(r, z', \omega), \tag{2.50}$$

where

$$\tilde{E}_{\text{h}}(r, z', \omega) = \hat{F}[E_{\text{h}}(r, z', t')], \tag{2.51}$$

[7] See Appendix D for details.

and

$$\tilde{P}(r, z', \omega) = \hat{F}[P(r, z', t')]. \tag{2.52}$$

The source term on the right-hand side of Eq. (2.50) describes the response of medium to the laser field and includes both linear and nonlinear terms. It is convenient to separate the polarization into linear and nonlinear components as: $\tilde{P}(r, z', \omega) = \chi^{(1)}(\omega)\tilde{E}_h(r, z', \omega) + \tilde{P}_{nl}(r, z', \omega)$, where the linear susceptibility $\chi^{(1)}(\omega)$ includes both linear dispersion and absorption through its real and imaginary parts, respectively. The nonlinear polarization term $\tilde{P}_{nl}(r, z', \omega)$ can be expressed as

$$\tilde{P}_{nl}(r, z', \omega) = \hat{F}\{[n_0 - n_e(r, z', t')]D(r, z', t')\}, \tag{2.53}$$

where the free electron density $n_e(r, z', t')$ is calculated from Eq. (2.47), and $D(r, z', t')$ is the single-atom induced dipole moment caused by the fundamental driving laser field.

The refractive index $n(\omega) = \sqrt{1 + \chi^{(1)}(\omega)/\varepsilon_0}$ [43] (which is valid only off resonance or for the small absorption) is related to atomic scattering factors by

$$n(\omega) = 1 - \delta_h(\omega) - i\beta_h(\omega) = 1 - \frac{1}{2\pi} n_0 r_0 \lambda^2 (f_1 + i f_2), \tag{2.54}$$

where r_0 is the classical electron radius, λ is the wavelength of the harmonic, n_0 is again the neutral atom density, and f_1 and f_2 are atomic scattering factors obtained from [47, 48]. Note that $\delta_h(\omega)$ and $\beta_h(\omega)$ account for dispersion and absorption of the medium on high harmonics, respectively. Finally Eq. (2.50) can be written as

$$\begin{aligned} \nabla_\perp^2 \tilde{E}_h(r, z', \omega) - \frac{2i\omega}{c}\frac{\partial \tilde{E}_h(r, z', \omega)}{\partial z'} - \frac{2\omega^2}{c^2}(\delta_h + i\beta_h)\tilde{E}_h(r, z', \omega) \\ = -\omega^2 \mu_0 \tilde{P}_{nl}(r, z', \omega), \end{aligned} \tag{2.55}$$

where the nonlinear polarization as the source of high-harmonic field is explicitly given. After the propagation in the medium, one can obtain near-field harmonics at the exit face of gas jet ($z' = z_{out}$).

2.3.3 *High-Harmonic Field of Aligned Molecules*

In general both the fundamental laser field and the high-harmonic field are modified when they copropagate through a macroscopic medium. If both pressure and laser intensity are low, the effects of dispersion, Kerr nonlinearity and plasma defocusing on the fundamental laser field can be neglected.[8] In other words, the source term in Eq. (2.39) can be taken as zero, and the fundamental field is not modified through the

[8] This is also true for laser propagation in a partially aligned molecular medium.

medium. Under these conditions the profile of fundamental laser field in space (in the vacuum) can be expressed in an analytical form.[9] For the harmonic field, dispersion and absorption effects of the medium, explicitly expressed as a dispersion-absorption term in Eq. (2.55) are not included when the gas pressure is low. These effects would become important for the high gas pressure. For molecular targets, I will only focus on the experiments carried out under the conditions of low laser intensity and low gas pressure, and only include induced dipoles for the generated harmonic field.

The 3-D Maxwell's wave equation for the high harmonics in a partially aligned molecular gas medium is [49, 50]

$$\nabla^2 E_{\mathrm{h}}^{\parallel}(r, z, t, \alpha) - \frac{1}{c}\frac{\partial^2 E_{\mathrm{h}}^{\parallel}(r, z, t, \alpha)}{\partial^2 t} = \mu_0 \frac{\partial^2 P_{\mathrm{nl}}^{\parallel}(r, z, t, \alpha)}{\partial^2 t}. \tag{2.56}$$

Here $E_{\mathrm{h}}^{\parallel}(r, z, t, \alpha)$ and $P_{\mathrm{nl}}^{\parallel}(r, z, t, \alpha)$ are the parallel components (with respect to the polarization direction of the generating laser) of harmonic electric field and the nonlinear polarization caused by fundamental laser, respectively. α is the pump-probe angle, i.e., the angle between aligning (pump) and generating (probe) laser polarizations.

The nonlinear polarization term can be expressed as

$$P_{\mathrm{nl}}^{\parallel}(r, z, t, \alpha) = [n_0 - n_{\mathrm{e}}(r, z, t, \alpha)] D^{\parallel,\mathrm{tot}}(r, z, t, \alpha), \tag{2.57}$$

where $n_0 - n_{\mathrm{e}}(r, z, t, \alpha)$ gives the density of remaining neutral molecules, and $D^{\parallel,\mathrm{tot}}(t, \alpha)$ is the parallel component of induced single-molecule dipole over a number of active electrons (including the effects from outermost and inner molecular orbitals).

Note that $\theta(\theta')$ and $\phi(\phi')$ are polar and azimuthal angles of the molecular axis in the frame attached to the pump (probe) laser field [20]. These angles are related by

$$\cos\theta = \cos\theta' \cos\alpha + \sin\theta' \sin\alpha \cos\phi'. \tag{2.58}$$

The alignment distribution in the "probe" frame at the fixed time delay between pump and probe lasers is

$$\rho(\theta', \phi', \alpha) = \rho[\theta(\theta', \phi', \alpha)]. \tag{2.59}$$

The free-electron density $n_{\mathrm{e}}(t, \alpha)$ in Eq. (2.57) can be calculated as follows

$$n_{\mathrm{e}}(t, \alpha) = \int_0^{2\pi} \int_0^{\pi} n_{\mathrm{e}}(t, \theta') \rho(\theta', \phi', \alpha) \sin\theta' d\theta' d\phi'. \tag{2.60}$$

Here $n_{\mathrm{e}}(t, \theta')$ is the alignment-dependent free-electron density, obtained from

[9] If laser beam can be considered as a Gaussian beam, its spatial and temporal dependence is given approximately in Appendix D.3.

$$n_e(t,\theta') = n_0\{1 - \exp[-\int_0^t \gamma(\tau,\theta')d\tau]\}, \tag{2.61}$$

where $\gamma(\tau,\theta')$ is the alignment-dependent ionization rate calculated by the MO-ADK theory [27] for each individual molecular orbital.

By going to a moving coordinate frame ($z' = z$ and $t' = t - z/c$) again, Eq. (2.56) can be written in the frequency domain as

$$\nabla_\perp^2 \tilde{E}_h^{\parallel}(r,z',\omega,\alpha) - \frac{2i\omega}{c}\frac{\partial \tilde{E}_h^{\parallel}(r,z',\omega,\alpha)}{\partial z'} = -\omega^2\mu_0 \tilde{P}_{nl}^{\parallel}(r,z',\omega,\alpha), \tag{2.62}$$

where

$$\tilde{E}_h^{\parallel}(r,z',\omega,\alpha) = \hat{F}[E_h^{\parallel}(r,z',t',\alpha)], \tag{2.63}$$

and

$$\tilde{P}_{nl}^{\parallel}(r,z',\omega,\alpha) = \hat{F}[P_{nl}^{\parallel}(r,z',t',\alpha)]. \tag{2.64}$$

After the propagation in the medium, one can obtain the parallel component of near-field harmonics on the exit face of gas jet ($z' = z_{out}$). For isotropically distributed molecules and partially aligned molecules with $\alpha = 0°$ or $90°$, by symmetry there is only the parallel harmonic component. For partially aligned molecules with other angles, the perpendicular component, which is usually much smaller, would appear, and the high harmonics would be elliptically polarized in general [51]. Generalization of Eq. (2.62) to the perpendicular component is straightforward. I focus on the parallel component in this thesis only.

Note that Eqs. (2.44), (2.55) and (2.62) are solved using a Crank-Nicholson routine for each value of ω. Typical parameters used in the calculations are $200 \sim 400$ grid points along the radial direction r and 400 grid points along the longitudinal direction z for a 1-mm wide gas jet.

In the calculation, induced dipole moments included in the nonlinear polarizations of Eqs. (2.53) and (2.57) are mostly obtained by the QRS theory in this thesis. For an atomic target, one can use Eq. (2.33) to calculate the single-atom induced dipole in the frequency domain, and then transfer it back to the time domain by a Fourier transform. For a linear molecular target, it can only be partially aligned if it is placed in a short laser field (or a pump laser). The intensity of aligning laser is usually weak and not tightly focused such that it can be assumed to be a constant within the gas medium. In other words, the degree of molecular alignment is not varied in the medium at one fixed pump-probe time delay. The averaged induced dipole from partially aligned molecules at each point in the gas medium is then obtained by coherently averaging the induced dipole moment of a fixed-in-space molecule in Eq. (2.36) over a molecular angular distribution, i.e.,

$$D^{\parallel,\mathrm{avg}}(\omega,\alpha) = \int_0^{2\pi}\int_0^{\pi} D^{\parallel}(\omega,\theta')\rho(\theta',\phi',\alpha)\sin\theta' d\theta' d\phi'. \tag{2.65}$$

Equation (2.65) is only for one particular molecular orbital.[10]

The total laser induced dipole over a number of active electrons is written as [52, 53]

$$D^{\parallel,\mathrm{tot}}(\omega,\alpha)=\sum_{j,n} D^{\parallel,\mathrm{avg}}_{j,n}(\omega,\alpha), \tag{2.66}$$

where index j refers to the different molecular orbital, and n is an index to account for the degeneracy in each molecular orbital. In the end, Eq. (2.66) is transferred into the time domain, and put back into Eq. (2.57).

2.4 Far-Field Harmonic Emission

Once high harmonics are emitted at the exit plane of an atomic or molecular (probably aligned) gas medium (called near-field harmonics), as shown in Fig. 2.1, they could propagate further in the vacuum until they are detected by the spectrometer. In this process, high harmonics may go through a slit, an iris or a pinhole, or be reflected by a mirror or more complicated optical system before they reach the detector (called far-field harmonics). In an axial-symmetric optical system, the complex electric field on the initial plane (near field) is related to the final plane (far field) by an $ABCD$ ray matrix, and AD-BC=1 for a lossless system. Here I only consider the simplest configuration shown in Fig. 2.1 without any additional optics (or within the free space propagation) between near and far fields, and A=1, B=1, C=0 and D=1 in the $ABCD$ matrix. According to the diffraction theory in the paraxial approximation, far-field harmonics can be obtained by using near-field harmonics through a Hankel

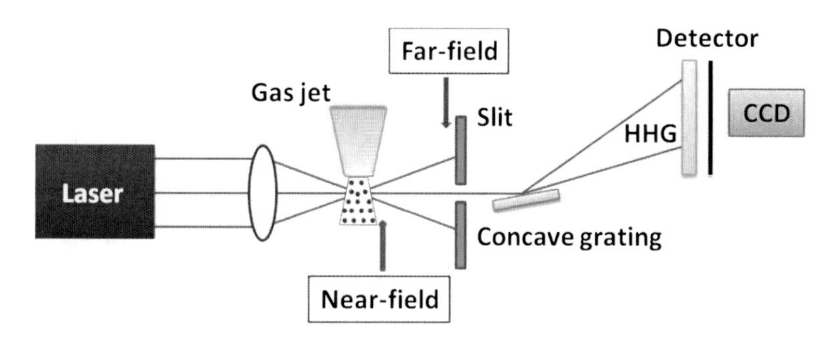

Fig. 2.1 Typical configuration for measuring the HHG in the far field. Adapted from [54]. © (2011) by IOP Publishing. Reproduced by permission of IOP Publishing. All rights reserved

[10] In [49, 50], the integral over ϕ' in Eqs. (2.60) and (2.65) is incorporated into $\rho(\theta',\alpha)$.

transformation [44, 55, 56][11, 12]

$$E_{\mathrm{h}}^{\mathrm{f}}(r_{\mathrm{f}}, z_{\mathrm{f}}, \omega) = -ik \int \frac{\tilde{E}_{\mathrm{h}}(r, z', \omega)}{z_{\mathrm{f}} - z'} J_0(\frac{krr_{\mathrm{f}}}{z_{\mathrm{f}} - z'}) \exp[\frac{ik(r^2 + r_{\mathrm{f}}^2)}{2(z_{\mathrm{f}} - z')}] r dr, \tag{2.67}$$

where J_0 is the zero-order Bessel function, z_{f} is the far-field position from the laser focus, r_{f} is the transverse coordinate in the far field, and the wave vector k is given by $k = \omega/c$. For a molecular target, $\tilde{E}_{\mathrm{h}}(r, z', \omega)$ is replaced by the electric field in Eq. (2.62), which also depends on the pump-probe angle α, and is parallel or perpendicular to the polarization of harmonic generating laser.

Assume that the high harmonics in the far field are collected from an extended area. By integrating the harmonic yield over this area, the power spectrum of macroscopic harmonics is obtained:

$$S_{\mathrm{h}}(\omega) \propto \int\int |E_{\mathrm{h}}^{\mathrm{f}}(x_{\mathrm{f}}, y_{\mathrm{f}}, z_{\mathrm{f}}, \omega)|^2 dx_{\mathrm{f}} dy_{\mathrm{f}}, \tag{2.68}$$

where x_{f} and y_{f} are Cartesian coordinates on the plane perpendicular to the propagation direction, and $r_{\mathrm{f}} = \sqrt{x_{\mathrm{f}}^2 + y_{\mathrm{f}}^2}$.

Note that in Eq. (2.68), the detailed information on the experimental setup is involved. To simulate experimental HHG spectra quantitatively, besides general used laser parameters, such as intensity, duration, wavelength, spot size, and so on, one needs more parameters about the experiment, for example, the size and location of a slit.

References

1. B.H. Bransden, C.J. Joachain, *Physics of Atoms and Molecules*, 2nd edn. (Prentice-Hall, New York, 2003)
2. C.J. Joachain, N.J. Kylstra, R.M. Potvliege, *Atoms in Intense Laser Fields* (Cambridge University Press, New York, 2012)
3. J.L. Krause, K.J. Schafer, K.C. Kulander, High-order harmonic generation from atoms and ions in the high intensity regime. Phys. Rev. Lett. **68**, 3535–3538 (1992)
4. K.C. Kulander, B.W. Shore, Calculations of multiple-harmonic conversion of 1064-nm radiation in Xe. Phys. Rev. Lett. **62**, 524–526 (1989)
5. J.C. Light, I.P. Hamilton, J.V. Lill, Generalized discrete variable approximation in quantum mechanics. J. Chem. Phys. **82**, 1400–1409 (1985)
6. X.M. Tong, S.I. Chu, Theoretical study of multiple high-order harmonic generation by intense ultrashort pulsed laser fields: A new generalized pseudospectral time-dependent method. Chem. Phys. **217**, 119–130 (1997)

[11] $\tilde{E}_{\mathrm{h}}(r, z', \omega)$ is expressed in the frequency domain, and its phase is also involved. Due to the different convention of Fourier transformation, this phase may need to change its sign before entering into the Eq. (2.67).

[12] Note that the elements of an $ABCD$ matrix are not expressed explicitly in Eq. (2.67), but the explicit expression can be found in Appendix D.2.

7. M. Lewenstein, Ph Balcou, M.Y. Ivanov, A. L'Huillier, P.B. Corkum, Theory of high-harmonic generation by low-frequency laser fields. Phys. Rev. A **49**, 2117–2132 (1994)
8. P.B. Corkum, Plasma perspective on strong field multiphoton ionization. Phys. Rev. Lett. **71**, 1994–1991 (1993)
9. X.X. Zhou, X.M. Tong, Z.X. Zhao, C.D. Lin, Role of molecular orbital symmetry on the alignment dependence of high-order harmonic generation with molecules. Phys. Rev. A **71**, 061801 (2005)
10. X.X. Zhou, X.M. Tong, Z.X. Zhao, C.D. Lin, Alignment dependence of high-order harmonic generation from N_2 and O_2 molecules in intense laser fields. Phys. Rev. A **72**, 033412 (2005)
11. C.C. Chirilă, M. Lein, Strong-field approximation for harmonic generation in diatomic molecules. Phys. Rev. A **73**, 023410 (2006)
12. J.P. Marangos, C. Altucci, R. Velotta, E. Heesel, E. Springate, M. Pascolini, L. Poletto, P. Villoresi, C. Vozzi, G. Sansone, M. Anscombe, J.-P. Caumes, S. Stagira, M. Nisoli, Molecular orbital dependence of high-order harmonic generation. J. Mod. Opt. **53**, 97–111 (2006)
13. A.T. Le, X.M. Tong, C.D. Lin, Evidence of two-center interference in high-order harmonic generation from CO_2. Phys. Rev. A **73**, 041402 (2006)
14. S. Ramakrishna, T. Seideman, Information content of high harmonics generated from aligned molecules. Phys. Rev. Lett. **99**, 113901 (2007)
15. S. Odžak, D. B. Milošević, Interference effects in high-order harmonic generation by homonuclear diatomic molecules. Phys. Rev. A, **79**, 023414 (2009)
16. A. Abdurrouf, F.H.M. Faisal, Theory of intense-field dynamic alignment and high-order harmonic generation from coherently rotating molecules and interpretation of intense-field ultrafast pump-probe experiments. Phys. Rev. A **79**, 023405 (2009)
17. P. Antoine, A. L'Huillier, M. Lewenstein, P. Salières, B. Carré, Theory of high-order harmonic generation by an elliptically polarized laser field. Phys. Rev. A **53**, 1725–1745 (1996)
18. M.J. Frisch et al., GAUSSIAN 03, Revision C.02 (Gaussian Inc., Pittsburgh, 2003)
19. T. Morishita, A.T. Le, Z. Chen, C.D. Lin, Accurate retrieval of structural information from laser-induced photoelectron and high-order harmonic spectra by few-cycle laser pulses. Phys. Rev. Lett. **100**, 013903 (2008)
20. A.T. Le, R.R. Lucchese, S. Tonzani, T. Morishita, C.D. Lin, Quantitative rescattering theory for high-order harmonic generation from molecules. Phys. Rev. A **80**, 013401 (2009)
21. C.D. Lin, A.T. Le, Z. Chen, T. Morishita, R. Lucchese, Strong-field rescattering physics–self-imaging of a molecule by its own electrons. J. Phys. B **43**, 122001 (2010)
22. A.T. Le, T. Morishita, C.D. Lin, Extraction of the species-dependent dipole amplitude and phase from high-order harmonic spectra in rare-gas atoms. Phys. Rev. A **78**, 023814 (2008)
23. H.G. Muller, Numerical simulation of high-order above-threshold-ionization enhancement in argon. Phys. Rev. A **60**, 1341–1350 (1999)
24. X.M. Tong, C.D. Lin, Empirical formula for static field ionization rates of atoms and molecules by lasers in the barrier-suppression regime. J. Phys. B **38**, 2593–2600 (2005)
25. R.R. Lucchese, G. Raseev, V. McKoy, Studies of differential and total photoionization cross sections of molecular nitrogen. Phys. Rev. A **25**, 2572–2587 (1982)
26. R.R. Lucchese, V. McKoy, Studies of differential and total photoionization cross sections of carbon dioxide. Phys. Rev. A **26**, 1406–1418 (1982)
27. X.M. Tong, Z.X. Zhao, C.D. Lin, Theory of molecular tunneling ionization. Phys. Rev. A **66**, 033402 (2002)
28. A.T. Le, R.D. Picca, P.D. Fainstein, D.A. Telnov, M. Lein, C.D. Lin, Theory of high-order harmonic generation from molecules with intense laser pulses. J. Phys. B **41**, 081002 (2008)
29. A.T. Le, R.R. Lucchese, M.T. lee, C.D. Lin, Probing molecular frame photoionization via laser generated high-order harmonics from aligned molecules. Phys. Rev. Lett. **102**, 203001 (2009)
30. A.T. Le, R.R. Lucchese, C.D. Lin, Uncovering multiple orbitals influence in high-harmonic generation from aligned N_2. J. Phys. B **42**, 211001 (2009)
31. E. Esarey, P. Sprangle, J. Krall, A. Ting, Self-focusing and guiding of short laser pulses in ionizing gases and plasmas. IEEE J. Quantum Electron. **33**, 1879–1914 (1997)

32. E. Takahashi, V. Tosa, Y. Nabekawa, K. Midorikawa, Experimental and theoretical analyses of a correlation between pump-pulse propagation and harmonic yield in a long-interaction medium. Phys. Rev. A **68**, 023808 (2003)
33. M. Geissler, G. Tempea, A. Scrinzi, M. Schnürer, F. Krausz, T. Brabec, Light propagation in field-ionizing media: Extreme nonlinear optics. Phys. Rev. Lett. **83**, 2930–2933 (1999)
34. M.B. Gaarde, J.L. Tate, K.J. Schafer, Macroscopic aspects of attosecond pulse generation. J. Phys. B **41**, 132001 (2008)
35. S.C. Rae, K. Burnett, Detailed simulations of plasma-induced spectral blueshifting. Phys. Rev. A **46**, 1084–1090 (1992)
36. E. Priori, G. Cerullo, M. Nisoli, S. Stagira, S. De Silvestri, P. Villoresi, L. Poletto, P. Ceccherini, C. Altucci, R. Bruzzese, C. de Lisio, Nonadiabatic three-dimensional model of high-order harmonic generation in the few-optical-cycle regime. Phys. Rev. A **61**, 063801 (2000)
37. M.V. Ammosov, N.B. Delone, V.P. Krainov, Tunnel ionization of complex atoms and of atomic ions in an alternating electromagnetic field. Sov. Phys. - JETP **64**, 1191–1194 (1986)
38. A. Börzsönyi, Z. Heiner, M.P. Kalashnikov, A.P. Kovács, K. Osvay, Dispersion measurement of inert gases and gas mixtures at 800 nm. Appl. Opt. **47**, 4856–4863 (2008)
39. P.J. Leonard, Dispersion measurement of inert gases and gas mixtures at 800 nm. At. Data Nucl. Data Tables **14**, 21–37 (1974)
40. H.J. Lehmeier, W. Leupacher, A. Penzkofer, Nonresonant third order hyperpolarizability of rare gases and N_2 determined by third harmonic generation. Opt. Commun. **56**, 67–72 (1985)
41. X.F. Li, A. L'Huillier, M. Ferray, L.A. Lompré, G. Mainfray, Multiple-harmonic generation in rare gases at high laser intensity. Phys. Rev. A **39**, 5751–5761 (1989)
42. J.K. Koga, N. Naumova, M. Kando, L.N. Tsintsadze, K. Nakajima, S.V. Bulanov, H. Dewa, H. Kotaki, T. Tajima, Fixed blueshift of high intensity short pulse lasers propagating in gas chambers. Phys. Plasmas **7**, 5223–5231 (2000)
43. R.W. Boyd, *Nonlinear Optics*, 2nd edn. (Academic Press, San Diego, 2003)
44. A. L'Huillier, Ph Balcou, S. Candel, K.J. Schafer, K.C. Kulander, Calculations of high-order harmonic-generation processes in xenon at 1064 nm. Phys. Rev. A **46**, 2778–2790 (1992)
45. H. Dachraoui, T. Auguste, A. Helmstedt, P. Bartz, M. Michelswirth, N. Mueller, W. Pfeiffer, P. Salières, U. Heinzmann, Interplay between absorption, dispersion and refraction in high-order harmonic generation. J. Phys. B **42**, 175402 (2009)
46. V. Tosa, H.T. Kim, I.J. Kim, C.H. Nam, High-order harmonic generation by chirped and self-guided femtosecond laser pulses. I. spatial and spectral analysis. Phys. Rev. A **71**, 063807 (2005)
47. C.T. Chantler, K. Olsen, R.A. Dragoset, J. Chang, A.R. Kishore, S.A. Kotochigova, D.S. Zucker, X-ray Form Factor, Attenuation and Scattering Tables (version 2.1). (National Institute of Standards and Technology, Gaithersburg, 2005)
48. B.L. Henke, E.M. Gullikson, J.C. Davis, X-ray interactions: Photoabsorption, scattering, transmission, and reflection at $E = 50$–$30{,}000$ eV, $Z = 1$–92. At. Data Nucl. Data Tables **54**, 181–342 (1993)
49. C. Jin, A.T. Le, C.D. Lin, Analysis of effects of macroscopic propagation and multiple molecular orbitals on the minimum in high-order harmonic generation of aligned CO_2. Phys. Rev. A **83**, 053409 (2011)
50. C. Jin, J.B. Bertrand, R.R. Lucchese, H.J. Wörner, P.B. Corkum, D.M. Villeneuve, A.T. Le, C.D. Lin, Intensity dependence of multiple-orbital contributions and shape resonance in high-order harmonic generation of aligned N_2 molecules. Phys. Rev. A **85**, 013405 (2012)
51. A.T. Le, R.R. Lucchese, C.D. Lin, Polarization and ellipticity of high-order harmonics from aligned molecules generated by linearly polarized intense laser pulses. Phys. Rev. A **82**, 023814 (2010)
52. C.B. Madsen, L.B. Madsen, High-order harmonic generation from arbitrarily oriented diatomic molecules including nuclear motion and field-free alignment. Phys. Rev. A **74**, 023403 (2006)
53. C. Figueira de Morisson Faria, B.B. Augstein, Molecular high-order harmonic generation with more than one active orbital: Quantum interference effects. Phys. Rev. A, **81**, 043409 (2010)

54. C. Jin, H.J. Wörner, V. Tosa, A.T. Le, J.B. Bertrand, R.R. Lucchese, P.B. Corkum, D.M. Villeneuve, C.D. Lin, Separation of target structure and medium propagation effects in high-harmonic generation. J. Phys. B **44**, 095601 (2011)
55. A.E. Siegman, *Lasers* (University Science, Mill Valley, 1986)
56. V. Tosa, K.T. Kim, C.H. Nam, Macroscopic generation of attosecond-pulse trains in strongly ionized media. Phys. Rev. A **79**, 043828 (2009)

Chapter 3
Medium Propagation Effects in High-Order Harmonic Generation of Ar

3.1 Introduction

High-order harmonic generation (HHG) is an extreme nonlinear optical process in which an intense ultrafast infrared laser light is efficiently converted to an ultrafast coherent extreme ultraviolet (XUV) or soft X-ray light. As discussed in Chap. 1, HHG has been widely studied for its potential as a short-wavelength light source [1], or the production of ultrashort light pulses [2]. It has also been shown to extract the information of atomic structure [3] or to image the molecular structure with sub-Angstrom precision in space and sub-femtosecond precision in time [4–7]. HHG process in single-atom response level can be intuitively understood in terms of the semiclassical "three-step" model [8, 9]. However, the laser field interacts with a macroscopic medium, and high harmonics from all atoms are generated coherently, a full description of experimentally observed harmonic spectra requires the treatment of the nonlinear propagation of fundamental laser beam together with high harmonics in the medium. As discussed in Chap. 2, the most accurate way to obtain the microscopic single-atom induced dipole is to solve the time-dependent Schrödinger equation (TDSE) numerically. Since this approach is quite time consuming and the calculations have to be carried out for hundreds of laser peak intensities in order to describe the nonuniform laser distribution inside a focused laser beam, this is rarely done in existing studies including the macroscopic propagation effect of HHG [10]. Instead, much simpler strong-field approximation (SFA) [11] is often used to calculate the single-atom induced dipole. Despite this limitation, the temporal and spatial properties of HHG observed experimentally have been reasonably understood from such SFA-based calculations. On the other hand, in a few examples, macroscopic HHG spectra obtained using TDSE-calculated induced dipoles do show significant quantitative discrepancies compared to SFA-based calculations [12, 13], and such studies have been limited to a few atomic gases only.

In this chapter, I demonstrate an accurate and efficient method for calculating the HHG spectrum from an atomic gaseous medium. This method is based on the recently developed quantitative rescattering (QRS) theory [14–16], which allows one

C. Jin, *Theory of Nonlinear Propagation of High Harmonics Generated in a Gaseous Medium*, Springer Theses, DOI: 10.1007/978-3-319-01625-2_3,

to calculate laser induced dipole of an atom with accuracy comparable to that obtained from solving the TDSE, yet with computing time comparable to that by using the SFA. The validity of QRS theory, at the single-atom level has been carefully calibrated against TDSE results for one-electron model atoms [17]. Clearly such comparison is incomplete without considering the phase-matching and macroscopic propagation effects. In this chapter, I first consider the situations where laser intensity and gas pressure are small such that the fundamental laser field is almost not modified during the propagation (or it can be assumed propagating in the vacuum). This simplification allows one to calculate macroscopic HHG spectra with TDSE-based single-atom induced dipoles, which can be used to calibrate the spectra with QRS-based induced dipoles. I then extend the theoretical model to more realistic situations of higher laser intensities and gas pressures, at which the nonlinear propagation of the fundamental field needs to be taken into account. I examine simulated HHG spectra of Ar and compare them directly with the experimental data. Based on the QRS theory, I show that the macroscopic HHG can be expressed as a product of a "macroscopic wave packet" (MWP) and a photorecombination (PR) cross section of the target. The MWP reflects the effect of laser and the consequence of its propagation in the medium.

In Sect. 3.2, with the single-atom response from TDSE or QRS, I will calculate macroscopic HHG spectra of Ar when a gas jet is put at the generally good phase-matching position. I will also show the phase of calculated macroscopic harmonics with changing gas-jet position. In Sect. 3.3, I will simulate HHG spectra of Ar with a 1200- or 1360-nm laser by taking into account the detailed experimental information, the experimental spectra from 30 to 90 eV can be accurately reproduced theoretically based on the QRS theory. In Sect. 3.4, I will show how width and depth of the well-known Cooper minimum in the HHG spectrum of Ar changes with gas-jet position. In Sect. 3.5, I will first show that the macroscopic HHG spectrum can be decomposed as a PR transition dipole (or a PR cross section) and a MWP. I will also verify that the MWP is a property of laser independent of the target by comparing the wave packets from different targets under the same laser. High harmonics are quite different by varying experimental conditions, and I find that all differences can be attributed to different MWP's. Thus I will show the dependence of MWP on the gas-jet position with respect to laser focus and on the gas pressure. Since phase-matching conditions are also dependent on the wavelength of laser used, I will investigate how the macroscopic HHG scales with laser wavelength in Sect. 3.6. In Sect. 3.7, I will give a summary of this chapter.

3.2 Macroscopic HHG Spectra: QRS Versus TDSE

In the numerical simulation, the fundamental laser pulse in space is taken to be a Gaussian beam with the cylindrical symmetry, propagating along z direction. The beam waist at the laser focus is fixed as $w_0 = 25$ μm, and a 1-mm long gas jet with a constant atomic density is placed after or at the laser focus. In time domain laser pulse is assumed to have a cosine-squared envelope, and carrier-envelope phase is taken to be $\varphi_{CE} = 0$ radian.[1]

In the SFA calculation, the ground-state wave function of Ar is obtained numerically by using the quantum chemistry software GAUSSIAN [18]. And for the QRS calculation, the model potential in Eq. (2.35) is used, which gives a Cooper minimum occurring near 42 eV instead of 51 eV from the Muller potential [19].

3.2.1 Strength of High Harmonics

By using the amplitude and phase of the single-atom induced dipole from TDSE, SFA and QRS as the source terms, one can calculate and compare the macroscopic HHG spectra from these three different models by solving the macroscopic propagation equation of high-harmonic field.[2]

In Fig. 3.1a, it shows single-atom HHG spectra of Ar, which are $\propto \omega^4 |D(\omega)|^2$ with the harmonic frequency ω and the induced dipole $D(\omega)$. The laser pulse has duration of 19.4 fs (full width at half maximum, FWHM), peak intensity of 1.5×10^{14} W/cm^2 and central wavelength of 800 nm. The spectra from QRS and SFA are normalized to that from the TDSE near the cutoff. One can see that HHG spectra in the plateau are very noisy, with no clear peaks at odd harmonics except in the cutoff region. It also shows that only for high harmonics close to the cutoff the SFA agrees with the TDSE, while in the plateau they show large discrepancies. For the QRS, on the other hand, except for a sharp spike near harmonic 14 (or H14), it agrees well with the TDSE. For the abnormal spike near H14, it can be easily traced to zero in the PR cross section of Ar in the plane wave approximation [20].

In Fig. 3.1b, macroscopic HHG spectra of Ar are shown when the gas jet with 1-mm length is placed 2 mm after the focus. Laser peak intensity at the center of the gas jet is 1.5×10^{14} W/cm^2. The same pulse duration and wavelength are applied as those in Fig. 3.1a. Assume that HHG signal after the propagation is collected right at the exit of the gas jet (near field). The macroscopic HHG spectra from QRS and SFA are also normalized to that from the TDSE in the cutoff. There are several general features presented in the spectra: well-resolved odd harmonics observed across the whole plateau; sharp drop of the spectra beyond the cutoff; smaller spectral widths in the plateau and their increases with the harmonic order; about the same cutoff location of the spectrum as that in single-atom response. The propagation greatly cleans up the spectrum between odd harmonics in comparison with the single-atom HHG spectrum. However, the relative intensity of odd harmonics does not change

[1] See Appendix D.3 for details.

[2] In the TDSE, the induced acceleration is first calculated in time domain and it is then converted to the induced dipole in frequency domain.

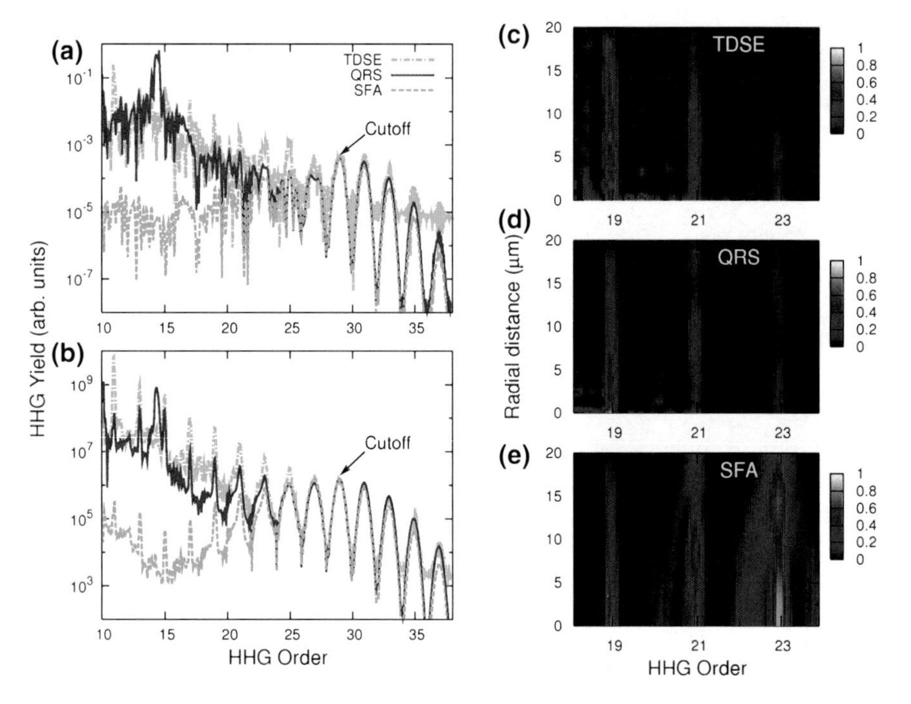

Fig. 3.1 **a** Single-atom harmonic spectra and **b** macroscopic harmonic spectra after the radial integration from TDSE, QRS and SFA. **c–e** Spatial distribution of the macroscopic harmonic emission at the exit face of gas jet. Atomic target: Ar. Adapted from [20]. © (2009) by the American Physical Society

too much after the propagation. At the cutoff, the SFA can give the correct prediction compared to the TDSE. But it obviously fails for the lower plateau spectrum. The QRS model, over the whole spectral region, gives a much better agreement with the one obtained from the TDSE after the propagation. Results in Fig. 3.1b also show that the QRS model, with the computational effort close to the SFA, is capable of improving the SFA quite significantly. Again the spike near H14 in the macroscopic spectra is caused by the same reason as in the single-atom case.

In Fig. 3.1b, it has only displayed the total HHG signal at the exit face of gas jet. It is interesting to investigate that in comparison with the TDSE how the QRS model improves over the SFA for the harmonic intensity at different radial points of exit face (which has the cylindrical symmetry). In Fig. 3.1c–e, they show the strength $|\widetilde{E}_h(r, z', \omega)|^2$ versus the radial distance for H19 to H23 based on three models. Again, there is the good overall agreement presented for TDSE and QRS. Their comparison also offers a good reason for adopting QRS-based single-atom induced dipoles instead of SFA-based ones for the high-harmonic propagation.

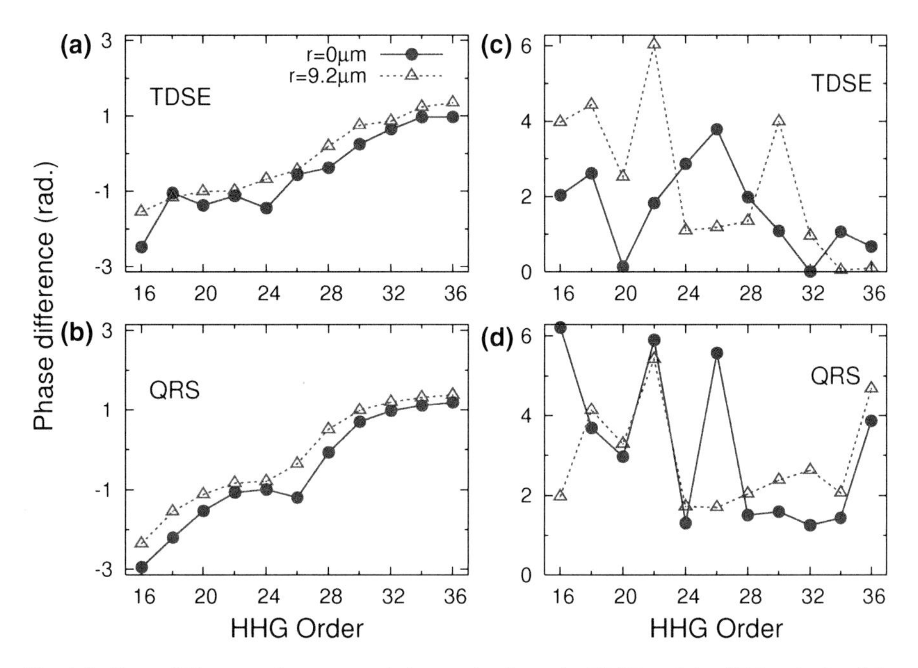

Fig. 3.2 Phase difference of macroscopic harmonics from the TDSE and the QRS, calculated for $r = 0$ μm or $r = 9.2$ μm at the exit plane of gas jet perpendicular to the propagation axis. **a** and **b** Gas jet is put 2 mm after the laser focus. **c** and **d** Gas jet is put at the laser focus. Laser duration and wavelength are the same as those in Fig. 3.1 while laser intensity at the center of gas jet is always kept as 1.5×10^{14} W/cm^2. Adapted from [20]. © (2009) by the American Physical Society

3.2.2 *Phase of High Harmonics*

Phases of high harmonics are very important for the attosecond pulse generation. In Eq. (2.31), it demonstrates that the phase of high harmonic has contributions from both the returning wave packet and the complex PR transition dipole moment. A following question is how the harmonic phase of single-atom response is affected by the macroscopic propagation. A proper way to present the phases of high harmonics is demanded to answer this question. The semiclassical theory tells that the harmonic emission time can be revealed by the phase difference between successive odd harmonics [21]. By using the RABITT technique [22], one could experimentally measure this phase difference between the consecutive odd harmonics. So the phase difference is defined by

$$\Delta\phi_{2n} = \phi_{2n+1} - \phi_{2n-1}. \tag{3.1}$$

In Fig. 3.2a and b, they show the phase difference of macroscopic HHG from the TDSE and the QRS when gas jet is placed after the focus, and the corresponding spectral strength has been shown in Fig. 3.1b. One can define the phase difference in the interval $[-\pi, \pi]$, successive phase differences at two different positions

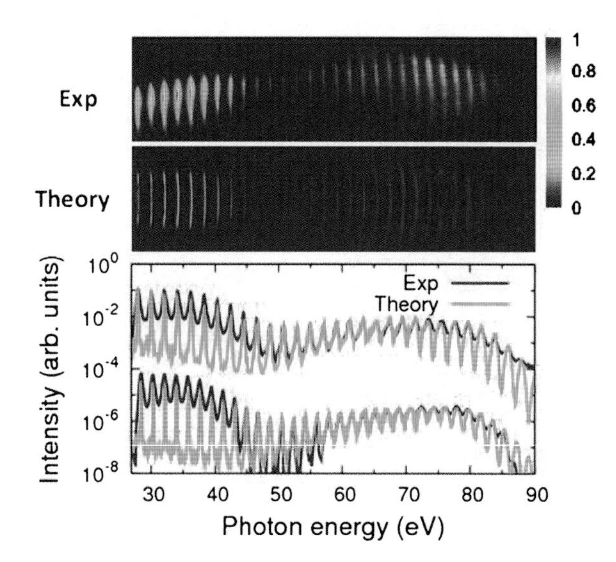

Fig. 3.3 High-harmonic spectra of Ar. *Upper frame*: Spatial distribution of harmonic emission versus photon energy in the far field by a 1200-nm laser. *Lower frame*: Comparison of experimental (*red lines*) and theoretical (*green lines*) HHG yields integrated over the vertical dimension for 1200-nm (*upper curves*) and 1360-nm (*lower curves*) lasers. Other laser parameters are given in the text. Adapted from [24].

$r = 0$ μm and $r = 9.2$ μm at the exit face of gas jet are shown. Because laser intensity changes rapidly radially, the harmonic fields emitted at varied radial positions would be quite different, it is meaningful to observe the phase behavior of high harmonics at different positions separately. Figure 3.2a shows that the phase difference increases almost linearly with harmonic order (linear chirp or attochirp [21]) with the almost same slope for both $r = 0$ μm and $r = 9.2$ μm due to the phase-matching, but the curve for $r = 0$ μm is shifted down in comparison with that for $r = 9.2$ μm. One can easily derive that the absolute phase increases quadratically with the harmonic order in these two cases. In Fig. 3.2b, the QRS demonstrates the same phase behavior as that in Fig. 3.2a for the TDSE. The capability of the QRS in studying the macroscopic response is shown here again.

In order to fully understand the mechanism of high-harmonic phase behavior after the macroscopic propagation, the gas jet is moved at the laser focus. Laser peak intensity at its center is fixed to be 1.5×10^{14} W/cm^2, and other laser parameters are kept the same as those in Fig. 3.2a and b. One can see the phase difference of macroscopic HHG from TDSE and QRS in Fig. 3.2c and d. Both TDSE and QRS show random-like phase differences, which are similar to the single-atom HHG, no matter $r = 0$ μm or $r = 9.2$ μm. Note that the observation of phase behavior here agrees with Gaarde et al.'s TDSE calculation [23], see their Fig. 3.

3.3 Macroscopic HHG Spectra: Theory Versus Experiment

To compare with the experimental HHG measurements, one should carry out the propagation for both fundamental and harmonic fields in the medium, and take into account the further propagation of high harmonics in the vacuum, i.e., the information of detecting system. The spatial beam mode of fundamental laser field at the entrance of gas medium is assumed as a Gaussian one. In the QRS calculation, the Muller potential in Eq. (2.34) is used to obtain the PR cross section of Ar, and the returning electron wave packet is obtained from the SFA, where Ar is assumed as a hydrogenlike atom in Eq. (2.30).

Figure 3.3 shows measured and simulated HHG spectra of Ar. Experimentally, a 0.5-mm-long gas jet was placed few mm's after laser focus. A vertical slit with a width of 100 μm was placed 24 cm after gas jet. For a 1200- (1360-) nm laser used in the experiment, the beam waist at the focus is 47.5 (52.5) μm, and the pulse duration is ~ 40 ($\sim$50) fs. Laser intensity and gas pressure in the simulations are adjusted until the best overall fit is achieved with the experimental data. For the 1200-nm laser, peak intensity for the experiment (theory) is $1.6\,(1.5)\times 10^{14}$ W/cm^2, and gas pressure is 28 (84) Torr. For the 1360-nm laser, the corresponding intensity and pressure are $1.25\ (1.15)\times 10^{14}$ W/cm^2 and 28 (56) Torr, respectively. In the upper frame of Fig. 3.3, the horizontal axis is photon energy, and the vertical axis is transverse spatial dimension. Experimental and theoretical spectra, generated by the 1200-nm laser, are normalized at harmonic 75 (H75), or at photon energy of 77 eV. One can see the general agreement between two spectra except for "up-down" asymmetry in the experimental spectra, which is due to the asymmetry in the laser beam profile. The "famous" Cooper minimum is clearly seen in both experimental and theoretical spectra. Harmonic yields integrated over the vertical dimension are compared in the lower frame in Fig. 3.3. The HHG spectrum with a 1360-nm laser is also shown. In both cases, one can see a good agreement (in the envelope of HHG spectrum) between theory and experiment over the 30–90 eV region covered. A careful examination of Fig. 3.3 reveals that there are still small discrepancies between experimental data and simulations by the QRS despite various attempts using somewhat different laser parameters. The harmonic width (or harmonic chirp) [25], mainly determined by laser intensity, pulse duration and gas pressure, is narrower in the simulation than that in the experimental measurement. In experiments, the parameters such as gas pressure, laser intensity and its spatial distribution cannot be measured precisely. Other factors, such as the use of slit and the position of detector, can also influence the HHG spectra. All of these uncertainties can contribute to the discrepancy between the simulation and the measured HHG spectrum.

3.4 Disappearance of Cooper Minimum in the HHG Spectrum of Ar

Cooper minimum (CM) in the HHG spectrum of Ar has been studied intensively by using traditional 800-nm lasers [26–28] and longer-wavelength lasers [29, 30]. This minimum always occurs in the single-atom HHG spectrum of Ar based on the QRS theory, however, it is not necessary to appear in the macroscopic spectrum. The position of CM can change or even disappear under different experimental

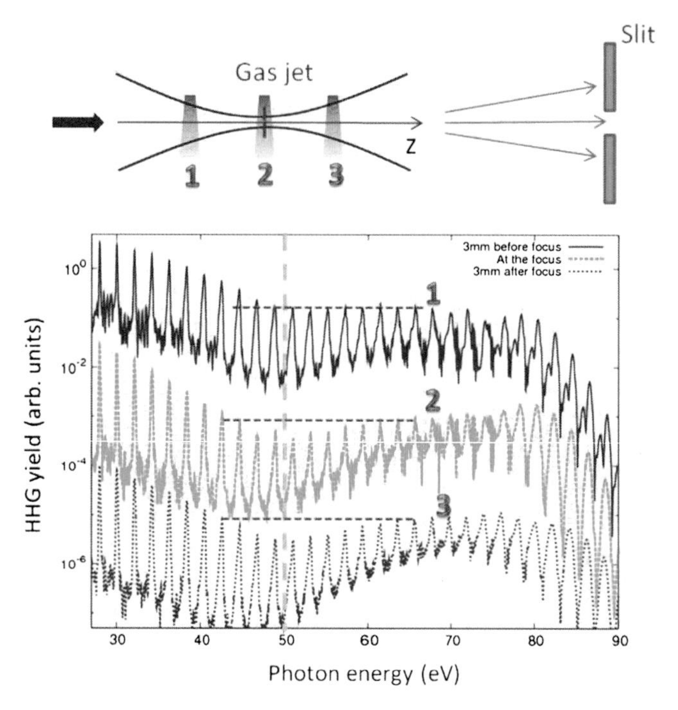

Fig. 3.4 Variation of width and depth of the Cooper minimum in the HHG spectrum of Ar using a 1200-nm laser. *Upper frame*: Sketch of the experimental setup where gas jet is put at different positions with respect to the laser focus. *Lower frame*: Calculated HHG spectra of Ar at three different gas-jet positions. *Yellow dashed line* indicates the position of Cooper minimum using the Muller potential [19]. Laser intensity in the gas jet is 1.6×10^{14} W/cm^2 and gas pressure is 56 Torr. Other parameters are the same as those in Fig. 3.3

conditions. Experiments have shown that the CM may disappear in the HHG spectrum by changing gas pressure [27] or by changing gas-jet position with respect to the laser focus [28].

To make this point clear, one can also carry out the calculations for different experimental conditions as shown in the upper frame of Fig. 3.4. In the lower frame of Fig. 3.4, calculated HHG spectra of Ar are shown in the far field with changing gas-jet position. In case 3, the CM is wide and deep in the spectrum. With moving the gas jet close to the laser focus, the CM becomes shallow in case 2, and high harmonics around 70 eV do not show clean peak structures indicating that the phase matching conditions become poor. If one places the gas jet before laser focus as shown in case 1, the phase matching conditions are greatly changed since the geometric phase of fundamental laser changes its sign. The CM disappears in the spectrum, and some of high-energy harmonics show noisy structures. This example once again tells that the CM in the macroscopic HHG spectrum can be washed out easily by changing the macroscopic experimental conditions. The change of experimental conditions can also be reflected by the "macroscopic wave packet", which will be discussed in Sect. 3.5.

3.5 Macroscopic Wave Packet

Based on the QRS theory, the macroscopic HHG spectrum in the near field or in the far field for an atomic target can be expressed as

$$S_{\mathrm{h}}(\omega) \propto \omega^4 |W'(\omega)|^2 |d(\omega)|^2, \tag{3.2}$$

where $W'(\omega)$ (the complex amplitude) is the "macroscopic wave packet" (MWP), and $d(\omega)$ is the PR transition dipole moment. The MWP can be considered as the collective effect of the microscopic wave packet for returning electrons, which is mostly governed by the fundamental laser field. And the MWP can be theoretically calculated by solving the Maxwell's propagation equation. In other words, laser, gas medium and experimental setup effects are all incorporated into the MWP, while the structure information of target is included in the PR transition dipole. In this chapter, only the amplitude of MWP is concerned.

3.5.1 Independence of Macroscopic Wave Packet on Targets

To extract the PR transition dipole (or PR cross section) from macroscopic HHG spectrum, one needs to rely on the MWP in Eq. (3.2). The question is how to obtain the MWP, which is only a property of laser and experimental setup. One can answer this question in a different way. Taking Ar target as an example, one can first carry out the propagation calculation to obtain the macroscopic HHG by using the QRS model to generate single-atom induced dipoles. In this calculation, the returning electron wave packet of single atom is obtained from the SFA by using the ground state wave function of Ar. The MWP can be extracted by using Eq. (3.2). In Fig. 3.5, it shows the resulting MWP $|W'(\omega)|$ of Ar (solid line) from the HHG spectra in Fig. 3.1b. Using laser parameters and focusing condition the same as those in Fig. 3.1b, the other MWP calculated in terms of a hydrogenlike atom is shown in Fig. 3.5, where the effective nuclear charge has been adjusted such that its $1s$ binding energy is the same as the $3p$ ground state energy of Ar. By normalizing two MWPs at the cutoff (marked by an arrow and estimated from laser peak intensity at the center of gas jet), one can see that they agree relatively well. These results indeed show that MWPs from different targets with the same I_p agree with each other reasonably well under the same laser condition.[3] These results have important implications. In general, the atomic PR transition dipole is well known. By measuring HHG spectra of an atomic target and a molecular one with nearly identical binding energy under the same laser field, one can extract the transition dipole of the molecule by calculating the ratio of HHG yields of two targets and using the known PR transition dipole of the atomic target. This model has been assumed in Itatani et al. [4] and in Levesque et al. [31]. Results here confirm the validity of their assumptions.

[3] More examples of the MWP can be found in [20].

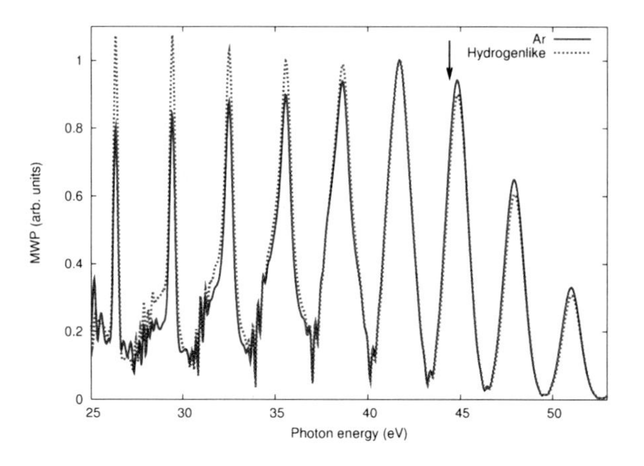

Fig. 3.5 Macroscopic wave packets extracted from macroscopic HHG spectra based on the QRS using Ar and hydrogenlike atom. Gas jet is placed 2 mm after the laser focus. Laser intensity is 1.5×10^{14} W/cm^2. *Arrow* indicates the cutoff position determined by the classical model. Adapted from [20]. © (2009) by the American Physical Society

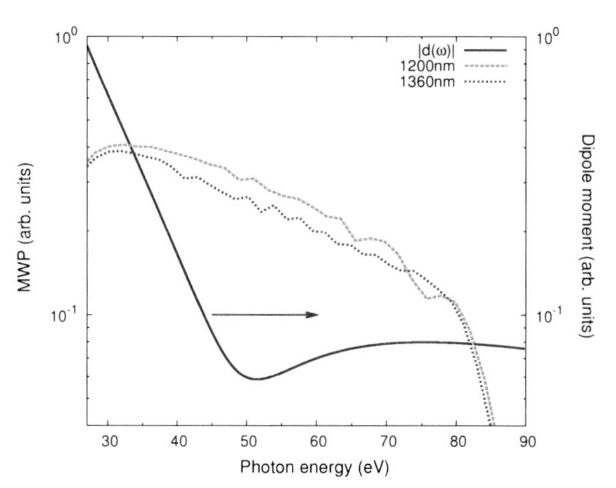

Fig. 3.6 Macroscopic wave packet extracted from HHG spectra of Ar for 1200- and 1360-nm lasers in Fig. 3.3. Calculated photorecombination transition dipole moment using the Muller potential [19] is also shown. Adapted from [24]. © (2011) by IOP Publishing. Reproduced by permission of IOP Publishing. All rights reserved

3.5.2 Separation of Target Structure Information from HHG Spectra

According to Eq. (3.2), HHG spectra in the lower frame of Fig. 3.3 can be decomposed as the MWPs and the PR transition dipole moment in Fig. 3.6. Two wave packets derived from 1200-nm to 1360-nm lasers are quite similar, but they only have slight different slopes near 50 eV. The PR transition dipole, however, shows a clear and broad CM around 50 eV. Thus, one can conclude that the broad CM in the HHG spectrum shown in Fig. 3.3 is caused by the minimum in the PR transition dipole (or PR cross section). The precise position of CM in the HHG spectrum is somewhat influenced by the MWP. As an example shown in Fig. 3.6, due to the difference in the MWP, the position of CM with the 1200-nm laser in Fig. 3.3 is slightly different from that with the 1360-nm laser. This could explain that the measured position of

CM of Ar varies from one laboratory to another due to the sensitivity of MWP to the experimental conditions.

3.5.3 Dependence of Macroscopic Wave Packet on Experimental Conditions

The MWP is a characteristic of fundamental laser, which doesn't depend on the PR transition dipole of the target. In other words, macroscopic conditions only affect the HHG through its modifications on the MWP. Thus, one can only study the macroscopic propagation effect on the HHG by investigating how the MWP varies with laser and experimental conditions.

In Fig. 3.7a, it shows MWPs for three gas-jet positions, which are at $z = -3$ mm (gas jet before laser focus), $z = 0$ mm (at) and $z = 3$ mm (after). Three curves are extracted from the HHG spectra in Fig. 3.4. For easy visualization, only the smoothed envelope of $|W'(\omega)|$ is shown. Among three curves, "after" curve is very flat since the good phase-matching is favored for this arrangement as the single-atom harmonic phase is partially compensated by the geometric phase of the focused laser. If gas jet is placed before the laser focus, the MWP changes rapidly, especially near the photon energy around 50 eV. Such strong energy dependence can wash out the CM in the HHG spectrum in Fig. 3.4.

In Fig. 3.7b it shows how the MWP depends on gas pressure for the focusing condition of $z = 3$ mm. Laser and other experimental parameters are the same as Fig. 3.7a. The MWP has been normalized by the ratio of pressure. Three curves would be on top of each other if a complete phase-matching condition had been fullfilled. The curve for higher pressure is slightly lower indicates that the full phase matching is not reached, especially for lower harmonics. With the increase of pressure, the MWP is much smoother versus energy. In fact, increasing the gas pressure

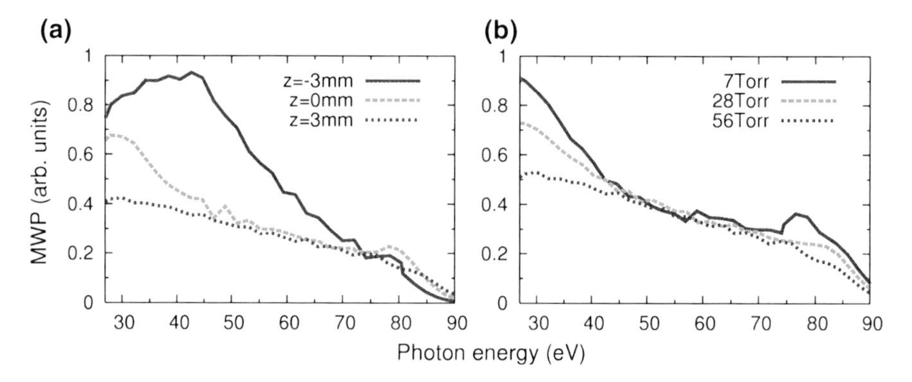

Fig. 3.7 **a** Dependence of macroscopic wave packet $|W'(\omega)|$ on the position of Ar gas jet with respect to the laser focus, and **b** on gas pressure. Adapted from [24].

tends to smooth out harmonics. These results also indicate that the harmonic energy increases quadratically with gas pressure, which is in agreement with the measurements reported in [32].

3.6 Wavelength Scaling of Harmonic Efficiency

One of main interests of studying the HHG is how to improve the harmonic efficiency, thus producing bright tabletop XUV or soft X-ray sources, or intense attosecond pulses. The single-atom harmonic energy at the cutoff is proportional to the square of driving laser wavelength, so high harmonics generated by near-infrared (NIR) or mid-infrared (MIR) lasers can efficiently reach high-energy photons, however, the yield is less favorable. How the HHG yield scales with laser wavelength becomes an interesting question. Only considering the single-atom response, there have been a few theoretical calculations [33–37]. However, macroscopic propagation effects in the medium have to be included in order to compare with the experimental HHG spectra. There have been a few investigations on the wavelength scaling of the HHG reported experimentally [29, 32, 38]. But the theoretical analysis including the phase-matching of high harmonics is still rather scarce.

To evaluate the harmonic yield at varied laser wavelength, one has to fix all parameters that may affect the HHG efficiency. It also needs to be determined if the total HHG yield or only the HHG yield within a given photon energy region is evaluated. Laser parameters in single-atom HHG simulations can be easily fixed, but it is more complex in the real experimental situation. So there are very few theoretical simulations including the macroscopic propagation effects [39, 40]. Because of the dependence of resulting HHG spectra on so many other parameters, any wavelength scaling laws derived are likely to depend on the experimental parameters used. Despite this limitation, one still can investigate the wavelength scaling by using the present QRS model. So a parameter that describes the efficiency of harmonic generation is first defined for this purpose. This parameter is the ratio of output energy (harmonic yield in a given energy region) with respect to input energy (pulse energy of the driving laser) at different laser wavelengths. The input pulse energy can be calculated by using Eq. (D.14) in Appendix D if the laser beam has a Gaussian distribution in time and space. The output energy can be easily obtained by integrating the harmonic intensity over a photon-energy region:

$$E_{\mathrm{out}} = \int_{\omega_{\mathrm{min}}}^{\omega_{\mathrm{max}}} \int |E_{\mathrm{h}}(x_f, y_f, \omega)|^2 dx_f dy_f d\omega. \tag{3.3}$$

The integration over the extended area is involved for the macroscopic HHG, where coordinates x_f and y_f are defined in Sect. 2.4.

Single-atom HHG spectra calculated at three wavelengths (800, 1200 and 1600 nm) are shown in Fig. 3.8a. To avoid noisy structures in the spectra, only the envelope of each spectrum is shown. For three wavelengths, laser intensity is kept as

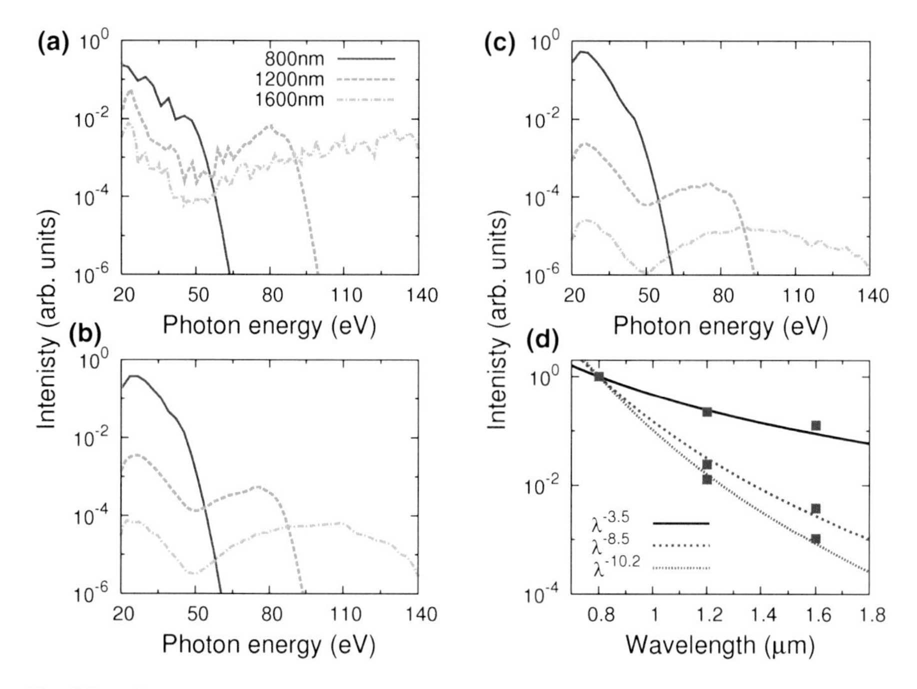

Fig. 3.8 **a** Single-atom HHG spectra, and macroscopic HHG spectra without **b** and with **c** using a slit for 800 nm (*solid lines*), 1200 nm (*dashed lines*) and 1600 nm (*dot-dashed lines*) lasers. Only the envelope of the spectrum is plotted in (**a**–**c**). Parameters used in the calculation are given in the text. **d** Wavelength dependence of integrated HHG yields above 20 eV. Integrated HHG yields in **a**, **b** and **c** follow scaling laws of $\lambda^{-3.5\pm0.5}$, $\lambda^{-8.5\pm0.5}$ and $\lambda^{-10.2\pm0.2}$, respectively. Adapted from [25]. © (2011) by the American Physical Society

1.6×10^{14} W/cm^2 while pulse duration is always 40 fs. In Fig. 3.8b it shows macroscopic HHG spectra obtained by including the phase-matching and propagation effects. In the calculation, beam waist at the focus is kept as 47.5 μm, and a 0.5 mm-long gas jet with gas pressure of 56 Torr is always placed at 3 mm after laser focus. By integrating the harmonic yield over the whole plane perpendicular to the propagation axis, one can obtain the total yield of each harmonic (without a slit). In Fig. 3.8c, HHG yields are recorded with the slit. High harmonics have diverged further in the vacuum and passed a slit (the slit with a width of 100 μm is placed 24 cm after the gas jet).

In single-atom simulations, the ratio of input energy is 1:1:1 for 800, 1200 and 1600 nm lasers in Fig. 3.8a. If one could define output energy as the integral of HHG yields above 20 eV, the resulting energy follows the scaling law of $\lambda^{-3.5\pm0.5}$ as shown in Fig. 3.8d. If one only integrates HHG yields between 20 eV and 50 eV, this would give a different scaling rule of λ^{-5}. In a previous study by Tate et al. [37], they fixed laser intensity and the number of optical cycles for 800 and 2000 nm lasers. So the ratio of input energy is 1:2.5 for 800 and 2000 nm lasers. And their scaling rules of $\lambda^{-(5-6)}$ with respect to a constant intensity would be modified as $\lambda^{-(6-7)}$ in terms of a constant input energy.

It is generally known [37, 41] that the phase-matching conditions are more difficult to fulfill if longer-wavelength lasers are applied. Thus the HHG efficiency decreases with the increase of laser wavelength. Here one can consider total HHG yields for lasers used in Fig. 3.8b where gas jet is placed at $z = 3$ mm after laser focus. Laser intensities are fixed at the center of the gas jet and laser beam waists at the focus are also fixed for three wavelengths. Laser focusing condition is governed by the wavelength only according to the diffraction theory. So the intensities at laser focus are 1.78×10^{14} W/cm^2, 2.01×10^{14} W/cm^2 and 2.33×10^{14} W/cm^2, for 800, 1200 and 1600 nm lasers, respectively. It can be easily calculated that input energies have the ratios of 1:1.13:1.31. As shown in Fig. 3.8d, HHG yields integrated from 20 eV up scale like $\lambda^{-8.5\pm0.5}$. Assume that only harmonic yields between 20 and 50 eV are integrated as the output energy, it gives the scaling rule of $\lambda^{-10.5}$.

In one experiment, Colosimo et al. [29] reported that HHG yields between 35 and 50 eV for an 800-nm laser are about 1000 times bigger than that for a 2000-nm laser, and experimental conditions have been kept "as fixed as possible". One can see a λ^{-9} dependence in this narrow energy region which is very close to the above scaling of $\lambda^{-10.5}$. In another experiment, Shiner et al. [32] demonstrated a scaling rule of $\lambda^{-6.3\pm1.1}$ for HHG of Xe with a fixed laser intensity, which was to be compared to the scaling law derived in the single-atom response. It is hard to compare their results with the scaling laws in this thesis due to some uncertainties in the experiment, such as spatial laser profile, position of the gas jet, gas pressure in the interaction region, and so on. So any simple scaling laws derived should be taken with caution of experimental parameters.

In Fig. 3.8d, it also shows the scaling law for HHG yields which are collected after a slit. By integrating HHG signals above 20 eV, one could obtain $\lambda^{-10.2\pm0.2}$ scaling. In general, the good phase-matching condition is difficult to meet if the long-wavelength laser is applied. Even if the gas jet is placed after the laser focus, both "short" and "long" trajectories (maybe a portion of "long" trajectories) could survive after the macroscopic propagation. The slit is usually used to select the contributions from "short" trajectories in the far field. Relative contributions from "short" and "long" trajectories are varied with the laser wavelength. By blocking out the contributions from "long" trajectories the harmonic efficiency becomes worse, see Fig. 3.8c.

The above analysis tells that under the same experimental conditions HHG yields for long-wavelength driving lasers appear quite unfavorable. On the other hand, for the practical purpose, high-harmonic yields with a long-wavelength laser could be generated with some optimized conditions experimentally. In Colosimo et al. [29], HHG yields between 35 and 50 eV generated by using a 2000-nm laser can be as high as 50 % of that from an 800-nm laser if the experimental conditions for two wavelengths were optimized independently. Furthermore, Chen et al. [42] demonstrated that it was possible to generate the HHG for long-wavelength lasers by implementing a hollow-core waveguide filled with much higher-pressure gas. They can thus achieve usable photon yields even in the water-window region. It is clear that additional theoretical analysis on the macroscopic HHG at long-wavelength driving lasers under different experimental conditions is desirable.

3.7 Conclusion

In the past two decades, HHG by infrared laser pulses with atoms has been widely investigated both experimentally and theoretically. Since the HHG is generated from a macroscopic medium, theoretically one needs to deal with both the microscopic laser-atom interaction, and the macroscopic propagation of laser and harmonic fields. In this chapter, I have first showed that the calculated macroscopic HHG spectrum obtained from QRS-based atomic dipoles was in much better agreement with the TDSE than that from the SFA. For the sake of the TDSE being carried out efficiently, in these comparisons I have only considered the conditions of low-intensity lasers and low-density gas medium where the fundamental field can be considered as propagating in the vacuum. Then I have extended the model to higher laser intensities and gas pressures at which the nonlinear propagation of the fundamental laser field was also taken into account with the inclusion of dispersion, plasma and Kerr effects, and simulated HHG spectra of Ar with 1200- and 1360-nm lasers by considering the detailed experimental information. Experimental HHG spectra have been successfully reproduced by the theory. The most pronounced structure in the measured spectrum of Ar, i.e., Cooper minimum, has also been reproduced. Specifically, I have investigated how the Cooper minimum was washed out by changing the experimental conditions.

I have showed that the macroscopic HHG spectrum can be expressed as a product of a MWP and a single-atom photorecombination transition dipole moment. The MWP has been shown to be largely independent of target if the ionization potential is nearly the same for two targets. The study of HHG spectra with macroscopic conditions can be simplified as the study of MWPs only, which can be easily obtained by solving the Maxwell's equation with SFA-based induced dipoles. The concept of MWP also implies that one can extract the photorecombination transition dipole of an unknown atom or molecule from one for which the photorecombination transition dipole moment is known by comparing their measured HHG spectra in the same laser pulse.

It is always an obstacle to improve the harmonic efficiency if the long-wavelength laser is applied to obtain high-energy photons. Tate et al. [37] suggested that the harmonic yield followed a $\lambda^{-(5-6)}$ scaling at constant intensity theoretically. However, this scaling law was obtained by the calculation of single-atom response. It is generally known that the macroscopic propagation of HHG makes this scaling law even worse. In this chapter, I have showed the wavelength scaling law in the single-atom response first based on the QRS theory. And then I have fixed the input pulse energy, and showed how this scaling law was varied with the macroscopic conditions. This study also implies that the scaling law can be improved by changing the experimental conditions, such as increasing the gas pressure, increasing the input energy, optimizing the detecting system, and so on.

References

1. A. Rundquist, C.G. Durfee III, Z. Chang, C. Herne, S. Backus, M.M. Murnane, H.C. Kapteyn, Phase-matched generation of coherent soft X-rays. Science **280**, 1412–1415 (1998)
2. F. Krausz, M. Ivanov, Attosecond physics. Rev. Mod. Phys. **81**, 163–234 (2009)
3. A.D. Shiner, B.E. Schmidt, C. Trallero-Herrero, H.J. Wörner, S. Patchkovskii, P.B. Corkum, J.-C. Kieffer, F. Légaré, D.M. Villeneuve, Probing collective multi-electron dynamics in xenon with high-harmonic spectroscopy. Nat. Phys. **7**, 464–467 (2011)
4. J. Itatani, J. Levesque, D. Zeidler, H. Niikura, H. Pépin, J.C. Kieffer, P.B. Corkum, D.M. Villeneuve, Tomographic imaging of molecular orbitals. Nature **432**, 867–871 (2004)
5. O. Smirnova, Y. Mairesse, S. Patchkovskii, N. Dudovich, D. Villeneuve, P. Corkum, M. Yu. Ivanov, High harmonic interferometry of multi-electron dynamics in molecules. Nature **460**, 972–977 (2009)
6. H.J. Wörner, J.B. Bertrand, D.V. Kartashov, P.B. Corkum, D.M. Villeneuve, Following a chemical reaction using high-harmonic interferometry. Nature **466**, 604–607 (2010)
7. V.-H. Le, A.-T. Le, R.-H. Xie, C.D. Lin, Theoretical analysis of dynamic chemical imaging with lasers using high-order harmonic generation. Phys. Rev. A **76**, 013414 (2007)
8. P.B. Corkum, Plasma perspective on strong field multiphoton ionization. Phys. Rev. Lett. **71**, 1994–1991 (1993)
9. J.L. Krause, K.J. Schafer, K.C. Kulander, High-order harmonic generation from atoms and ions in the high intensity regime. Phys. Rev. Lett. **68**, 3535–3538 (1992)
10. M.B. Gaarde, M. Murakami, R. Kienberger, Spatial separation of large dynamical blueshift and harmonic generation. Phys. Rev. A **74**, 053401 (2006)
11. M. Lewenstein, Ph Balcou, M.Y. Ivanov, A. L'Huillier, P.B. Corkum, Theory of high-harmonic generation by low-frequency laser fields. Phys. Rev. A **49**, 2117–2132 (1994)
12. M.B. Gaarde, K.J. Schafer, Quantum path distributions for high-order harmonics in rare gas atoms. Phys. Rev. A **65**, 031406 (2002)
13. M.B. Gaarde, J.L. Tate, K.J. Schafer, Macroscopic aspects of attosecond pulse generation. J. Phys. B **41**, 132001 (2008)
14. C.D. Lin, A.T. Le, Z. Chen, T. Morishita, R. Lucchese, Strong-field rescattering physics–self-imaging of a molecule by its own electrons. J. Phys. B **43**, 122001 (2010)
15. T. Morishita, A.T. Le, Z. Chen, C.D. Lin, Accurate retrieval of structural information from laser-induced photoelectron and high-order harmonic spectra by few-cycle laser pulses. Phys. Rev. Lett. **100**, 013903 (2008)
16. A.T. Le, R.R. Lucchese, S. Tonzani, T. Morishita, C.D. Lin, Quantitative rescattering theory for high-order harmonic generation from molecules. Phys. Rev. A **80**, 013401 (2009)
17. A.T. Le, T. Morishita, C.D. Lin, Extraction of the species-dependent dipole amplitude and phase from high-order harmonic spectra in rare-gas atoms. Phys. Rev. A **78**, 023814 (2008)
18. M.J. Frisch et al., GAUSSIAN 03, Revision C.02 (Gaussian Inc., Pittsburgh, 2003)
19. H.G. Muller, Numerical simulation of high-order above-threshold-ionization enhancement in argon. Phys. Rev. A **60**, 1341–1350 (1999)
20. C. Jin, A.T. Le, C.D. Lin, Retrieval of target photorecombination cross sections from high-order harmonics generated in a macroscopic medium. Phys. Rev. A **79**, 053413 (2009)
21. Y. Mairesse, A. de Bohan, L.J. Frasinski, H. Merdji, L.C. Dinu, P. Monchicourt, P. Breger, M. Kovačev, R. Taieb, B. Carre, H.G. Muller, P. Agostini, P. Salières, Attosecond synchronization of high-harmonic soft X-rays. Science **302**, 1540–1543 (2003)
22. P.M. Paul, E.S. Toma, P. Breger, G. Mullot, F. Augé, Ph Balcou, H.G. Muller, P. Agostini, Observation of a train of attosecond pulses from high harmonic generation. Science **292**, 1689–1692 (2001)
23. M.B. Gaarde, K.J. Schafer, Space-time considerations in the phase locking of high harmonics. Phys. Rev. Lett. **89**, 213901 (2002)
24. C. Jin, H.J. Wörner, V. Tosa, A.T. Le, J.B. Bertrand, R.R. Lucchese, P.B. Corkum, D.M. Villeneuve, C.D. Lin, Separation of target structure and medium propagation effects in high-harmonic generation. J. Phys. B **44**, 095601 (2011)

25. C. Jin, A.T. Le, C.D. Lin, Medium propagation effects in high-order harmonic generation of Ar and N_2. Phys. Rev. A **83**, 023411 (2011)
26. H.J. Wörner, H. Niikura, J.B. Bertrand, P.B. Corkum, D.M. Villeneuve, Observation of electronic structure minima in high-harmonic generation. Phys. Rev. Lett. **102**, 103901 (2009)
27. S. Minemoto, T. Umegaki, Y. Oguchi, T. Morishita, A.T. Le, S. Watanabe, H. Sakai, Retrieving photorecombination cross sections of atoms from high-order harmonic spectra. Phys. Rev. A **78**, 061402 (2008)
28. J.P. Farrell, L.S. Spector, B.K. McFarland, P.H. Bucksbaum, M. Gühr, M.B. Gaarde, K.J. Schafer, Influence of phase matching on the Cooper minimum in Ar high-order harmonic spectra. Phys. Rev. A **83**, 023420 (2011)
29. P. Colosimo, G. Doumy, C.I. Blaga, J. Wheeler, C. Hauri, F. Catoire, J. Tate, R. Chirla, A.M. March, G.G. Paulus, H.G. Muller, P. Agostini, L.F. DiMauro, Scaling strong-field interactions towards the classical limit. Nat. Phys. **4**, 386–389 (2008)
30. J. Higuet, H. Ruf, N. Thiré, R. Cireasa, E. Constant, E. Cormier, D. Descamps, E. Mével, S. Petit, B. Pons, Y. Mairesse, B. Fabre, High-order harmonic spectroscopy of the Cooper minimum in argon: experimental and theoretical study. Phys. Rev. A **83**, 053401 (2011)
31. J. Levesque, D. Zeidler, J.P. Marangos, P.B. Corkum, D.M. Villeneuve, High harmonic generation and the role of atomic orbital wave functions. Phys. Rev. Lett. **98**, 183903 (2007)
32. A.D. Shiner, C. Trallero-Herrero, N. Kajumba, H.-C. Bandulet, D. Comtois, F. Légaré, M. Giguère, J.-C. Kieffer, P.B. Corkum, D.M. Villeneuve, Wavelength scaling of high harmonic generation efficiency. Phys. Rev. Lett. **103**, 073902 (2009)
33. M.V. Frolov, N.L. Manakov, A.F. Starace, Wavelength scaling of high-harmonic yield: threshold phenomena and bound state symmetry dependence. Phys. Rev. Lett. **100**, 173001 (2008)
34. P. Lan, E.J. Takahashi, K. Midorikawa, Wavelength scaling of efficient high-order harmonic generation by two-color infrared laser fields. Phys. Rev. A **81**, 061802 (2010)
35. A. Gordon, F. Kärtner, Scaling of keV HHG photon yield with drive wavelength. Opt. Express **13**, 2941–2947 (2005)
36. K. Schiessl, K.L. Ishikawa, E. Persson, J. Burgdörfer, Quantum path interference in the wavelength dependence of high-harmonic generation. Phys. Rev. Lett. **99**, 253903 (2007)
37. J. Tate, T. Auguste, H.G. Muller, P. Salières, P. Agostini, L.F. DiMauro, Scaling of wave-packet dynamics in an intense midinfrared field. Phys. Rev. Lett. **98**, 013901 (2007)
38. T. Popmintchev, M.-C. Chen, O. Cohen, M.E. Grisham, J.J. Rocca, M.M. Murnane, H.C. Kapteyn, Extended phase matching of high harmonics driven by mid-infrared light. Opt. Lett. **33**, 2128–2130 (2008)
39. V.S. Yakovlev, M. Ivanov, F. Krausz, Enhanced phase-matching for generation of soft X-ray harmonics and attosecond pulses in atomic gases. Opt. Express **15**, 15351–15364 (2007)
40. E.L. Falcão-Filho, V.M. Gkortsas, A. Gordon, and F. X. Kärtner, Analytic scaling analysis of high harmonic generation conversion efficiency. Opt. Express **17**, 11217–11229 (2009)
41. B. Shan, Z. Chang, Dramatic extension of the high-order harmonic cutoff by using a long-wavelength driving field. Phys. Rev. A **65**, 11804 (2001)
42. M.-C. Chen, P. Arpin, T. Popmintchev, M. Gerrity, B. Zhang, M. Seaberg, D. Popmintchev, M.M. Murnane, H.C. Kapteyn, Bright, coherent, ultrafast soft X-ray harmonics spanning the water window from a tabletop light source. Phys. Rev. Lett. **105**, 173901 (2010)

Chapter 4
Comparison of High-Order Harmonic Generation of Ar Using a Truncated Bessel or a Gaussian Beam

4.1 Introduction

As discussed in Chap. 3, a full quantitative description of high-order harmonic generation (HHG) in a macroscopic medium requires the inclusion of the propagation of fundamental laser field and generated harmonic field. The QRS theory has been successfully incorporated into the well-established macroscopic propagation theory such that simulated HHG spectra can be compared directly with the experimental measurements in Fig. 3.3, where the experimental conditions have been well specified. High harmonics in these studies were generated with multi-cycle (FWHM, $\sim$10 optical cycles) laser pulses. And these simulations were based on the assumption that the initial fundamental laser pulse at the entrance of gas medium was a Gaussian beam. Few-cycle laser pulses are also widely used to produce high harmonics, and they are usually obtained by gas-filled hollow-core fiber compression technique [1]. In this method, an incident laser beam can be dominantly coupled into the fundamental EH_{11} hybrid mode by proper mode matching. At the exit of the fiber a truncated Bessel (TB) beam is produced instead of a Gaussian beam. Nisoli et al. [2] have shown that using a TB beam as the driving laser pulse the spatial properties (divergence and brightness) of high harmonics were greatly improved. To simulate high harmonics generated by few-cycle pulses, the macroscopic propagation code is generalized to include the conditions where the spatial distribution of generating laser pulse is a TB beam.

In this chapter, using the spatial TB beam I first want to check if the high-harmonic spectra of Ar reported by Wörner et al. [3] can be simulated, which were performed with a few-cycle laser pulse at relatively high intensities (Fig. 1 of their paper). The second goal of this chapter is to establish the general conditions where the generated harmonic spectra are not sensitive to whether the generating beam is a Gaussian beam or a TB beam.

The numerical propagation code is modified by changing the initial condition at the entrance of a gas jet in which the input beam is a TB beam. A TB beam exiting from the hollow-core fiber is usually propagated and diverged further in the vacuum,

C. Jin, *Theory of Nonlinear Propagation of High Harmonics Generated in a Gaseous Medium*, Springer Theses, DOI: 10.1007/978-3-319-01625-2_4,

and refocused through lens and mirrors before entering the harmonic-generating gas medium. In the Appendix D.2, two types of TB beams are described. In TB-1 (Type-1 Bessel), it shows a tight focusing beam used by Nisoli et al. [2]. In TB-2 (Type-2 Bessel), a loosely focused TB beam is shown, which was used by Wörner et al. [3] and Shiner et al. [4]. In Sect. 4.2 I will show calculated HHG spectra of Ar with either a TB beam or a Gaussian beam. A 780-nm laser and the setup parameters as close as those in Wörner et al.'s [3] are used in the calculations. Even with the TB-2 beam, observed deep Cooper minimum (CM) in the Wörner et al.'s experiment still cannot be reproduced. However, I am able to show that HHG spectrum of Ar reported in Shiner et al.'s [4] is reproduced. High-harmonic spectra in their experiment were generated by using 1800-nm mid-infrared lasers. I then turn to study the detailed harmonic emission in space for TB-1 and TB-2 beams in terms of the phase matching conditions. In Sect. 4.3, I will investigate how these harmonic growth maps change with gas-jet position. In Sect. 4.4, I will specifically study the pressure induced phase mismatch by analyzing the harmonic growth map for different gas pressures. And then the conclusion is drawn that HHG spectra produced from a TB-2 beam are generally close to those generated from a Gaussian beam with the similar beam waist. A summary in Sect. 4.5 will conclude this chapter.

4.2 Simulations of HHG Spectra of Ar

4.2.1 780-nm Few-Cycle Laser

Photoionization cross section (PICS) of Ar has a pronounced minimum [6] at photon energy near 51 eV. This is well-known as the Cooper minimum (CM). This Cooper minimum appearing in the harmonic spectrum of Ar has been reported in many measurements [3, 4, 7–10] under different experimental conditions (different laser intensities, laser wavelengths or laser beam profiles). In order to observe CM in the harmonic spectrum clearly, the harmonic cutoff should go well beyond 51 eV. With typical 800-nm Ti: sapphire lasers, this would require a high laser intensity because the harmonic cutoff energy is proportional to the pondermotive energy. However, laser intensity is limited due to the ground-state depletion. Thus the CM in the HHG spectrum of Ar was not clearly located in earlier experiments with 800-nm lasers. Wörner et al. [3] used a few-cycle laser pulse to avoid the saturation effect, and they observed clear CM. In their experiment, a laser pulse from a hollow-core fiber filled with Ar gas used to achieve self-phase modulation, was compressed to a few-cycle pulse ($\sim$3 optical cycles) by subsequently using chirped mirrors. The appearance of a clear deep CM at 53 ± 3 eV was the most prominent feature in this experiment, which did not shift with laser intensity. The QRS theory (in the singe-atom level) generally predicts that the position of CM in the HHG spectrum is at about 51 eV, and it does not change with laser intensity. Wörner et al.'s [3] measurements appear to be consistent with this prediction qualitatively. However, the width and depth of the CM

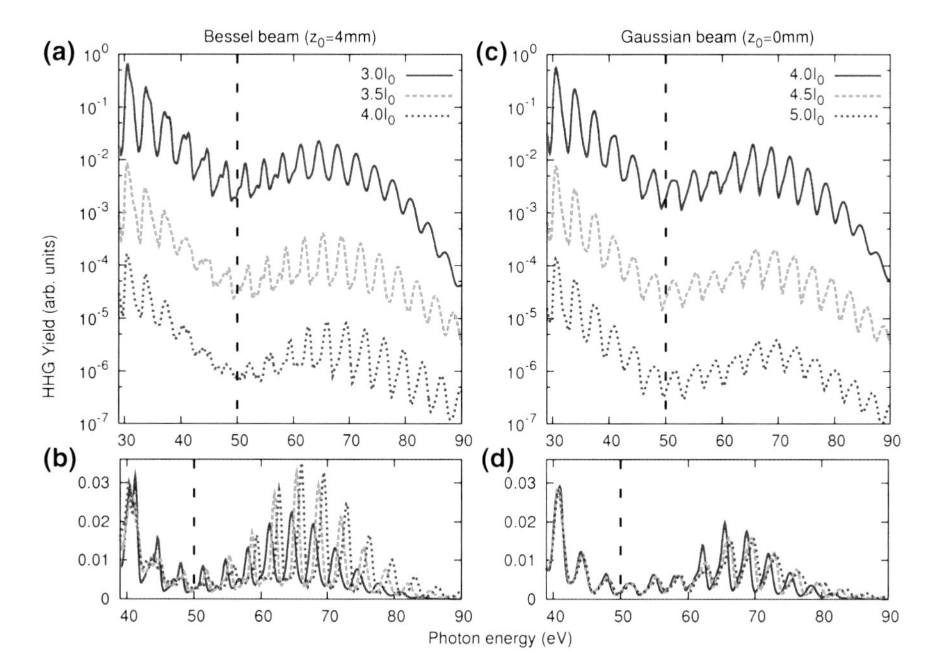

Fig. 4.1 Simulated HHG spectra (after CEP averaged) of Ar as the initial laser profile is a truncated Bessel beam (TB-2) (**a** and **b**), or a Gaussian beam (beam waist $w_0 = 50\,\mu\text{m}$) (**c** and **d**). z_0 is the gas-jet position with respect to laser focus, and laser intensity (at the focus, $z = 0$ mm) is given in units of $I_0 = 10^{14}$ W/cm^2. *Dashed lines* indicate the Cooper minimum position. Laser wavelength is 780 nm. See text for additional laser parameters. Adapted from [5]. © (2012) by the American Physical Society

appear to contradict the QRS prediction quantitatively. Note that CM observed in the PICS of Ar is not so deep [6]. Assuming that incident laser pulses were Gaussian beams, I have carried out the simulations with the experimental parameters, but was unable to reproduce the broad and deep Cooper minimum reported in the experiment, see Fig. 4.1. Is this probably due to the use of a Gaussian beam in the simulation? To answer this question, I will show results from the simulations using a truncated Bessel beam.

In the simulation, both the fundamental and harmonic field are propagated in the medium, and the single-atom induced dipole is obtained by using the QRS theory. Calculation parameters are chosen close to those in Wörner et al. [3]. Laser wavelength is 780 nm, and duration is three cycles (FWHM). Gas jet is 1-mm wide in the interaction region and gas pressure is assumed to be a constant at 30 Torr, and a slit with a width of 100 μm is placed 24 cm after the gas jet to select high harmonics in the far field. To obtain the correct experimental cut-off position, laser peak intensity at the focus (in the vacuum) was adjusted as indicated in Fig. 4.1.

I first assume that the incident beam is a TB-2 pulse, and the center of gas jet is located at 4 mm after laser focus (i.e., $z_0 = 4$ mm). HHG spectra after CEP averaged are

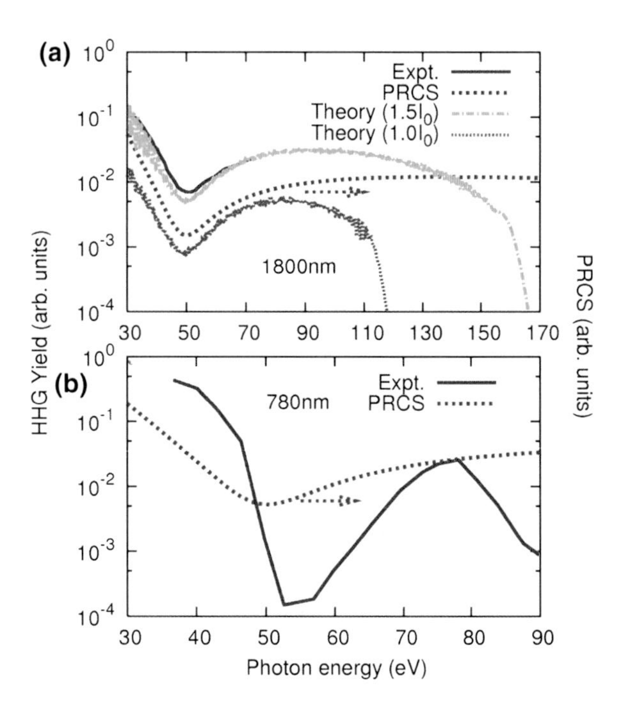

Fig. 4.2 **a** Comparison of experimental (envelope only) [4] and theoretical HHG spectra (after CEP averaged) using an 1800-nm laser. Laser intensities used in the simulations are indicated in units of $I_0 = 10^{14}$ W/cm^2. See text for additional parameters. Experimental data are shown only from 30 to 75 eV due to constraint from the filter. **b** Experimental HHG spectra (envelope only) [3] by a 780-nm laser with the intensity of 2.9×10^{14} W/cm^2. This plot can be compared with Fig. 4.1. Calculated differential photorecombination cross section (PRCS) using the Muller potential [11] is also shown in **a** and **b**. Adapted from [5]. © (2012) by the American Physical Society

shown in Fig. 4.1a. For clarity, the spectra have been shifted for different intensities. In Fig. 4.1b, I show the spectra for three intensities in linear scale. Laser intensities indicated are ones at the laser focus ($z = 0$ mm), so on-axis laser intensities at $z = 4$ mm are almost the same as those in Gaussian pulses (at $z = 0$ mm) in Fig. 4.1c. I next assume that laser pulse is a Gaussian beam with the waist $w_0 = 50\,\mu$m and the center of gas jet is at the laser focus ($z_0 = 0$ mm). HHG spectra after CEP averaged are shown in Fig. 4.1c. I also show the spectra in linear scale in Fig. 4.1d. High-harmonic spectra shown in Fig. 4.1a do not differ significantly from those in Fig. 4.1c, with the CM appearing near 50 eV. When the laser field is high enough and reaches the saturation, higher harmonics show blue shift. In Fig. 4.1c, the ratio of maximum yield near the cutoff with respect to the lowest yield at the CM is about a factor of 3–6 in the simulation, but the same ratio is close to 100 in the experiment of Wörner et al. [3], also see Fig. 4.2b below. I have varied the position of gas jet (z_0) with respect to laser focus in the case of Gaussian beam, but high-harmonic spectra remain nearly the same as those in Fig. 4.1c. From Fig. 4.1b, I find that with the TB-2 beam, HHG spectra are stronger for the higher harmonics, such that the previous maximum/minimum ratio rises by about 50 %, but still much smaller than the ratio seen in Wörner et al. [3]. I have also changed z_0 (not shown) in the case of TB-2 beam, the CM was always seen, but the depth of CM reported in the experiment still cannot be reproduced. Thus the origin of discrepancy remains unexplained.

4.2.2 1800-nm Few-Cycle Laser

As shown in Eq. (1.4), the cutoff energy can be extended with the increase of laser wavelength since the ponderomotive energy is proportional to the square of wavelength of a driving laser. It is preferable to study CM in the HHG spectrum of Ar using near-infrared (NIR) lasers. Indeed, such measurements have been reported by Higuet et al. [10] with 1.8–2.0-μm and 50-fs NIR lasers and by Jin et al. [12] using 1.2- and 1.36-μm lasers. None of the CMs in the HHG spectra reported in these experiments are as deep as shown in Wörner et al. [3]. In fact, the simulations using an incident Gaussian beam spatially can well reproduce the experimental data reported in Jin et al. [12]. Recently, Shiner et al. [4] also reported HHG spectra measurements of Ar using a few-cycle (~2 optical cycles) 1800-nm laser (see Fig. 9 in the supplementary information). Their experimental spectrum is shown in Fig. 4.2a. I choose the calculation parameters as close as those in Shiner et al. [4]. The simulation is carried out with an 1800-nm and 11-fs laser pulse. The initial laser beam is assumed as a Gaussian one with $w_0 = 100\,\mu$m. A gas jet (0.5 mm wide) is located at the laser focus, gas pressure is 6 Torr, and a slit with a width of 190 μm is placed at 45.5 cm after the gas jet. Only high harmonics after the slit are detected. Calculated HHG spectra (after CEP averaged) from 30 to 170 eV with two intensities are shown in Fig. 4.2a. One can see that the experimental spectrum in the given energy region agrees very well with the theoretical one (laser intensity is 1.5×10^{14} W/cm^2). In the low photon energy region, these spectra also agree well with the calculated PRCS of Ar using the Muller potential [11]. If laser intensity is decreased (at 1.0×10^{14} W/cm^2), except that the cut-off position moves to lower photon energy, the general spectral shape and the depth of Cooper minimum don't change much. This also shows that the experiment of Shiner et al. [4] can be modeled by using a Gaussian beam.

Based on above simulations, it is concluded that the deep Cooper minimum in the HHG spectrum reported in Wörner et al. [3] (as shown in Fig. 4.2b) remains not reproduced by the theory even with including the phase-matching and propagation effects. On the other hand, this deep minimum was not observed in other experiments using NIR lasers [4] while the simulations can reproduce these latter observations.

4.3 Phase Matching Conditions at Low Gas Pressure

In a previous study, Nisoli et al. [2] have shown that resulting HHG characteristics using a truncated Bessel beam as an initial beam was quite different from that using a Gaussian beam. Results in Sect. 4.2 seem not consistent with their conclusions. It turns out that in Sect. 4.2, a loosely focused truncated Bessel beam (or TB-2) was applied, while in Nisoli et al. [2] they used a tightly focused truncated Bessel beam (or TB-1). In Appendix D.2 two types of TB beams (TB-1 and TB-2) are constructed, their typical spatial intensity distributions are also presented. Here I will present a systematic comparison of the phase-matching conditions of high harmonics produced by TB-1, TB-2 and Gaussian beams. In the calculation, *ab initio* macroscopic propagation is applied, single-atom induced dipole is obtained by the QRS theory. Laser intensity at the focus, wavelength, duration (FWHM) and CEP are fixed at 3×10^{14} W/cm^2, 780 nm, 3 optical cycles and 0, respectively.[1]

As discussed in Sect. 1.3.1, the phase matching is a pre-requisite for efficient generation of high harmonics. The phase mismatch for q-th harmonic in Eq. (1.5) mainly includes four terms [13–17]. Each term has been discussed in detail in Sect. 1.3.1. Here I only give a brief review. In Eq. (1.5), the first term is due to laser focusing (or geometry phase), the second term is from free-electron dispersion (or plasma dispersion), the third term is from neutral atom dispersion where the index of refraction changes with wavelength, and the last term is caused by laser-induced atomic dipole phase depending strongly on laser intensity. The phase mismatch of induced dipole is given by $K_{q,dip} = \nabla\varphi_{q,dip}$, where $\varphi_{q,dip}$ is the classical action accumulated by an electron during its excursion in the laser field, which could follow either "long" or "short" trajectory.

4.3.1 Phase Matching Map at Low Gas Pressure

Gas pressure is a very important factor to adjust the conditions of high harmonics. I first set the gas pressure very low (0.1 Torr) such that the pressure effect can be ignored for the time being. In this case, phase-matching is only determined by the interplay between the geometric phase $\varphi_{q,geo}(r,z)$ and the induced dipole phase $\varphi_{q,dip}(r,z)$ in Eq. (1.8). I plot $\Delta\varphi_q(r,z) = \varphi_{q,geo}(r,z) - \varphi_{q,dip}(r,z)$, modulo 2π, in Fig. 4.3 for 15th harmonic (H15). This plot is in contrast to generally used contour map of the coherence length [13, 18]. The color coding is chosen such that it is dark (or red) when $\Delta\varphi_q$ is near 0 and 2π, and bright (or white) when $\Delta\varphi_q$ is near π such that there are no color changes at two boundaries. Note that the length scale is in mm along z-axis while it is in μm along r-axis. The phase difference between two neighboring white regions (or red regions) is 2π in Fig. 4.3. If the white (or red) region is large, the corresponding gradient is small, and the phase matching is good. One can see in general from this figure that it is more difficult to achieve the good phase matching for "long" (lower row) than for "short" (upper row) trajectories.

One can calculate the phase mismatch by taking the gradient of the phase in Fig. 4.3. The calculated phase mismatch $\Delta k_q(z)$ along the propagation axis z for H15 generated by a Gaussian beam is shown in Fig. 4.4c and d. The magnitude of the phase mismatch for "short"-trajectory is much smaller than that for "long"-trajectory one. Furthermore, one can see that the phase matching is better after the focus (for positive z values). Laser intensity decreases quickly away from the laser focus, thus a gas jet located at $z = 2$ mm is about the optimum condition for H15 using a Gaussian beam. Figure 4.4a and b show the phase mismatch $\Delta k_q(z)$ for a TB beam along the propagation axis z. Again the magnitude of phase mismatch is much larger for "long"-

[1] The distributions of laser intensity and phase in space are plotted in Figs. D.3 and D.5.

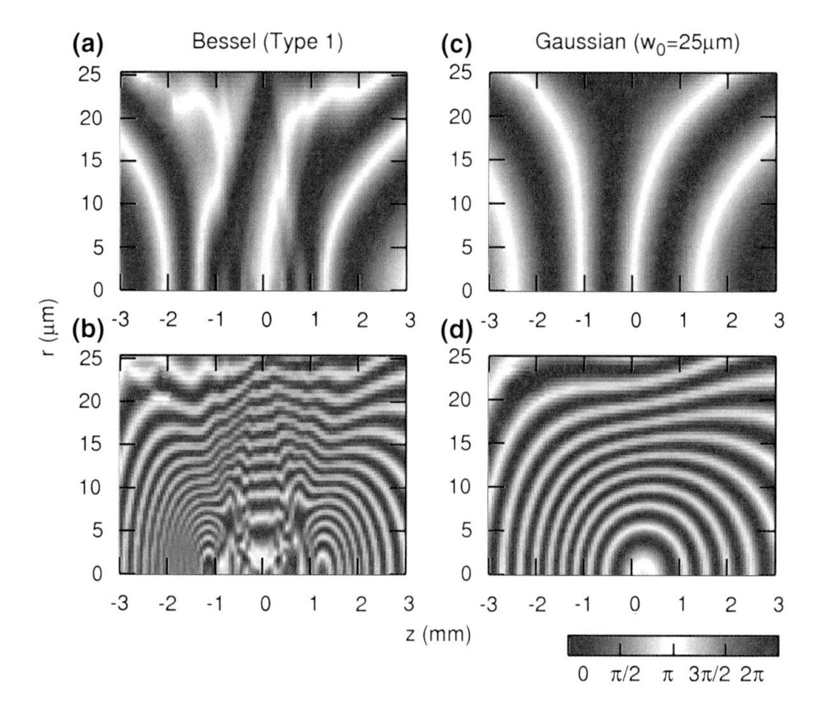

Fig. 4.3 Interplay of the phase matching for H15 between the geometric phase $\varphi_{q,geo}(r, z)$ and the induced dipole phase $\varphi_{q,dip}(r, z)$ for Type-1 Bessel and Gaussian ($w_0 = 25\,\mu$m) beams. *Upper row*: "short" trajectory; *lower row*: "long" trajectory. Note that $\Delta\varphi_q(r, z) = \varphi_{q,geo}(r, z)$-$\varphi_{q,dip}(r, z)$ modulo 2π is plotted, and the phase change between two neighboring *white regions* (or two neighboring *red regions*) is 2π. Adapted from [5]. © (2012) by the American Physical Society

trajectory than that for "short"-trajectory one. Although the phase oscillates widely near $z = 0$ mm, it would result in a small phase mismatch by the spatial average over a small volume. Thus for TB beams a broad good phase-matching region can be achieved close to the axis from $z = -1.5$ to 1.5 mm for "short"-trajectory harmonics. This argument is actually consistent with the conclusion in Nisoli et al. [2]. These figures tell that phase-matching conditions depend strongly on the position of gas jet with its typical length ($\sim$1 mm) no matter it is a Type-1 Bessel or Gaussian ($w_0 = 25\,\mu$m) beam.

For off-axis harmonics, there are components of the phase mismatch parallel and perpendicular to the axis. For a Gaussian beam, the distance between two white regions (or two red regions) where the phase changes by 2π, is larger along z-axis than along r-axis, see Fig. 4.3c and d, thus the phase matching (by taking the gradient of phase) is still favorable off axis, even not as good as the on-axis region (also see Fig. 4 in [13]). In Fig. 4.3, it also shows that, in general, "long"-trajectory harmonics tend to be phase matched off axis, and thus they are more divergent.

I next consider the loosely focused laser beams. As shown by Eqs. (D.4) and (D.5) in Appendix D, for a Gaussian beam, if one can scale r by the beam waist w_0

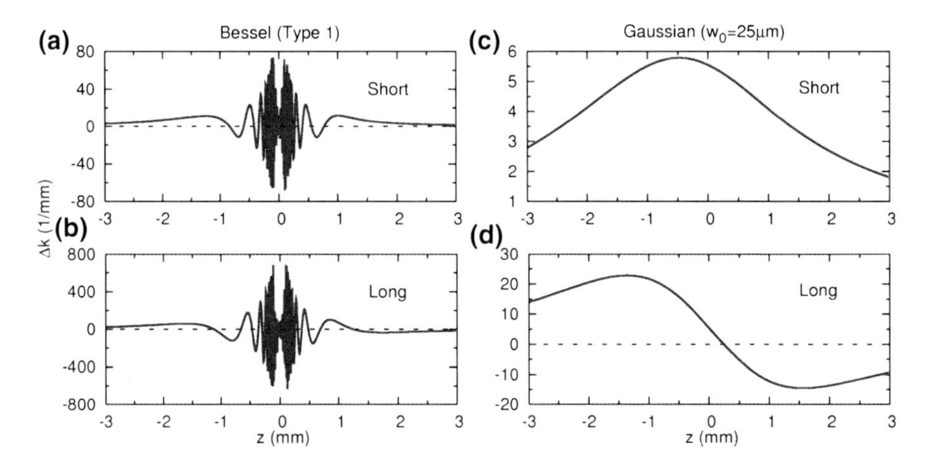

Fig. 4.4 On-axis phase mismatch $\Delta k_q(z) = (\partial/\partial z)[\Delta\varphi_q(0, z)]$ for the phase shown in Fig. 4.3 with Type-1 Bessel and Gaussian ($w_0 = 25\,\mu$m) beams. $q = 15$. *Upper row*: "short" trajectory; *lower row*: "long" trajectory. Zero values of the phase mismatch are indicated by *dashed lines*. Note that values of $\Delta k_q(z)$ from $z = -0.2$ to 0.2 mm in **a** and **b** are probably not precise numerically due to the dramatic phase oscillation along z direction. Adapted from [5]. © (2012) by the American Physical Society

and scale z by the confocal parameter b, laser intensity and phase don't change in the scaled coordinate. Thus the phase-matched volume will increase scaled by b or w_0 in each direction for the loosely focused Gaussian beam. In this case, one can expect that good phase-matching conditions are more easily achieved for a typical gas-jet length. This is also true for TB beams, and can only be verified numerically. I have checked (not shown) that the phase map for TB-2 beam was similar to (but not exactly the same as) that for TB-1 beam in scaled coordinates (for H15) since TB-1 and TB-2 beams are constructed differently. Thus for loosely focused TB-2 beams the phase-matched volume will also increase and phase matching conditions do not differ much from the loosely focused Gaussian beam for a typical gas-jet geometry. In Sect. 4.3.2, I will show that HHG spectra generated by TB-2 beam and Gaussian beam are very similar.

4.3.2 Dependence of Harmonic Yield on Gas-Jet Position

In Fig. 4.5, it shows spatial intensity distributions of the plateau harmonic (H15) and the cutoff harmonic (H35). High harmonics are generated by a TB-1 beam and a Gaussian beam under tight-focusing conditions at two different gas-jet (1-mm width) positions. To understand these results, I examine phase-mismatch values in units of 1/mm, see Eq. (1.5). On-axis phase mismatch $\Delta k_{q,geo}(0, z)$ and $K_{q,dip}(0, z)$ for a Gaussian beam are expressed as [14, 15]

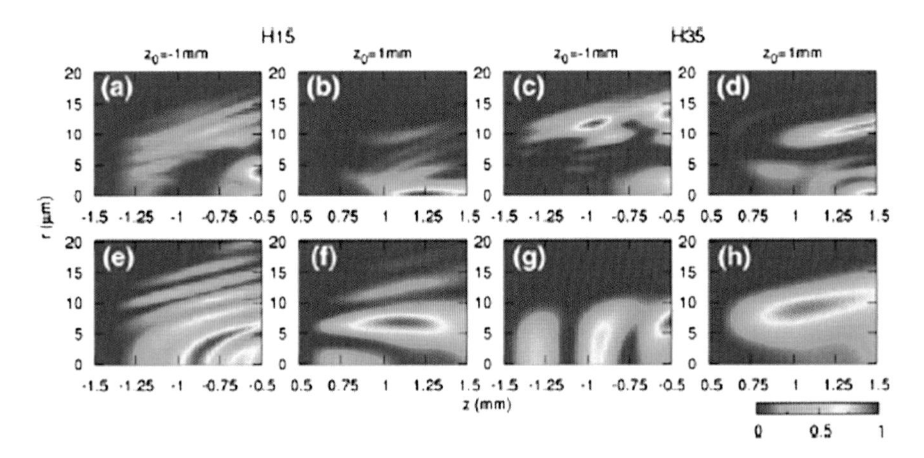

Fig. 4.5 Spatial distributions (normalized) of harmonic intensity for H15 and H35 using tight focusing laser beams. *Upper row*: Type-1 Bessel beam; *lower row*: Gaussian beam ($w_0 = 25\,\mu$m). z_0 is the position of gas-jet center with respect to laser focus, i.e., $z_0 > 0$ means the gas-jet center (1-mm wide) is placed after laser focus. Gas pressure: 0.1 Torr. Fundamental laser field is not modified through propagating in the medium at this pressure. Adapted from [5]. © (2012) by the American Physical Society

$$\Delta k_{q,geo}(0,z) \approx \frac{2}{b}(q-1)\frac{1}{1+(2z/b)^2}, \tag{4.1}$$

$$K_{q,dip}(0,z) = \frac{8z}{b^2}\frac{1}{[1+(2z/b)^2]^2}\alpha_i^q I_0. \tag{4.2}$$

Here α_i^q is defined in Eq. (1.8), and other parameters can be found in Appendix D. In Table 4.1, I show typical values of $\Delta k_{q,geo}(0,z)$ and $K_{q,dip}(0,z)$ calculated by using Eqs. (4.1) and (4.2) at $z = 1$ mm on the axis ($r = 0$). Note that the value of $K_{q,dip}(0,z)$ calculated by using α_i^q in the cut-off region in Eq. (1.8) may not be very accurate. For low gas pressure (0.1 Torr), there is no defocusing of the fundamental laser pulse. It is clearly shown in Table 4.1 that the "short" trajectory favors the good phase matching for $b = 5$ mm ($w_0 = 25\,\mu$m). The coherence length is defined as $l_{coh} = \pi/\Delta k_q$, where $\Delta k_q = \Delta k_{q,geo} - K_{q,dip}$. It is calculated to be about 1 mm for both H15 and H35. As

Table 4.1 Phase mismatch $\Delta k_{q,geo}(0,z)$ and $K_{q,dip}(0,z)$ (mm^{-1}) derived from Eqs. (1.6) and (1.8) for a Gaussian beam

Harmonic order		H15				H35			
Confocal parameter b (mm)		5	3	20	15	5	3	20	15
$\Delta k_{q,geo}$		4.83	6.46	1.39	1.83	11.72	15.69	3.37	4.45
$K_{q,dip}$	Short ("S")	0.71	1.28	0.059	0.103	<9.77	<17.51	<0.81	<1.41
	Long ("L")	17.12	30.67	1.41	2.47	>9.77	>17.51	>0.81	>1.41

Here $z = 1$ mm and $I_0 = 3 \times 10^{14}$ W/cm^2

seen in Fig. 4.5f and h, the steady growth of harmonic intensity along the propagation axis z is allowed by this large coherence length. If gas jet is placed before laser focus ($z_0 = -1$ mm), $K_{q,dip}$ changes its sign while $\Delta k_{q,geo}$ remains the same as that at $z = 1$ mm. No matter "short" or "long" trajectory, the coherence length l_{coh} becomes much smaller. In Fig. 4.5e and g, they show that the harmonic buildup along z is not monotonic. The small coherence length results in destructive interference such that the harmonic yield vanishes, followed by the buildup and then the destruction, as z increases. Thus gas-jet position $z_0 = -1$ mm is not favorable for the phase matching of high harmonics. In Fig. 4.5a–d, the harmonic spatial distribution for Type-1 Bessel beam is quite different from that for Gaussian beam, but the strong gas-jet position dependence is similar, i.e., the coherence length is longer for positive z_0 than that for negative z_0.

One can do the same analysis for a loosely focused Gaussian beam ($b = 20$ mm, $w_0 = 50\,\mu$m), and calculate l_{coh} by using the major phase-mismatch terms. l_{coh} is $\sim$2 mm (H15) or $\sim$1 mm (H35) at $z = 1$ mm, and it is $\sim$1 mm for both harmonics at $z = -2$ mm. The large coherence length allows steadily monotonic buildup of high harmonics as z is increased, as confirmed by numerical results shown in Fig. 4.6e–h. For Type-2 Bessel, as shown in Fig. 4.6a–d, the harmonic spatial distribution is very similar to the loosely focused Gaussian beam. These plots confirm that the harmonic spectra generated by loosely focused Gaussian and Bessel beams are expected to be quite similar at the same gas-jet position, and the harmonic spectra by either one of these beams are less sensitive to gas-jet positions as discussed in Sect. 4.2.

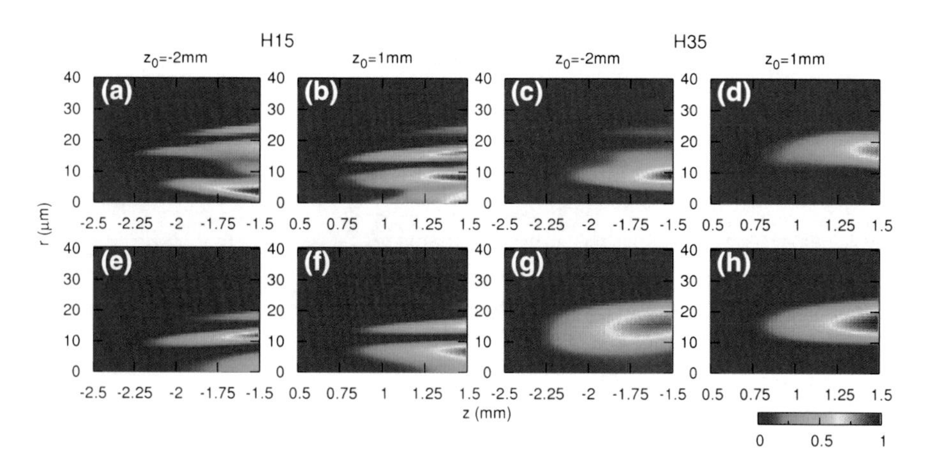

Fig. 4.6 Same as Fig. 4.5 except for loosely focused laser beams. *Upper row*: Type-2 Bessel beam; *lower row*: Gaussian beam ($w_0 = 50\,\mu$m). Adapted from [5]. © (2012) by the American Physical Society

4.4 Gas Pressure Induced Phase Mismatch

As mentioned previously gas pressure is crucial for phase matching conditions. Its effect cannot be neglected at high pressure. There are two dominant terms: the phase mismatch $\Delta k_{q,el}$ in Eq. (1.9) due to free electrons and $\Delta k_{q,at}$ in Eq. (1.10) due to neutral atom dispersion explicitly depending on the pressure [19]. $\Delta k_{q,el}$ is always positive, and $\Delta k_{q,at}$ usually is negative for high-energy photons in the XUV or soft X-ray. These two terms can be totally compensated, i.e., added up to zero, at relatively low ionization level (about 6 % for H15 and 4 % for H35) if a 780-nm and 3-cycle (FWHM) laser is applied. On the other hand, the highly free-electron density (due to high gas pressure) in the medium can also induce laser defocusing and blue shift, thus the geometric phase mismatch $\Delta k_{q,geo}$ in Eq. (1.6) and $K_{q,dip}$ in Eq. (1.7) are changed correspondingly. Variations of these values are difficult to quantify since the laser field undergoes complicated spatial and temporal variation in the medium. Actually it is still possible to quantify these values numerically even though there is no simple analytical form. In the following, I only illustrate the effect of laser defocusing by changing the confocal parameter b for a Gaussian beam, the effect of blue shift is not demonstrated separately.

At first I would like to give a rough estimate of two kinds of the phase mismatch caused by gas pressure, i.e., $\Delta k_{q,el} + \Delta k_{q,at}$ and $\Delta k_{q,geo} - K_{q,dip}$ due to laser defocusing. For tight focusing Gaussian beam, ionization level (in the end of the laser pulse) is about 12 %. So the values of $\Delta k_{q,el} + \Delta k_{q,at}$ are about 0.5 and 2 mm^{-1} at 10 Torr for H15 and H35, respectively. These values proportional to gas pressure increase to 4 and 16 mm^{-1} at 80 Torr. On the other hand, the high pressure leads to much free electron which could induce laser defocusing, i.e., making the confocal parameter b smaller. In Table 4.1, I show the values of $\Delta k_{q,geo}$ and $K_{q,dip}$ as b changes to 3 mm. One can see, at 10 Torr, the phase mismatch caused by laser focusing (mostly considering "short" trajectory only since $z = 1$ mm) is dominant, and then becomes comparable to $\Delta k_{q,el} + \Delta k_{q,at}$ at 80 Torr. For Type-1 Bessel and Gaussian beams, their geometric phase and induced dipole phase have been shown to be quite different, see Fig. 4.3. With the increase in gas pressure, harmonic spatial distributions for two beams behave differently because the phase differences between them still prevail at higher pressure. Figure 4.7 shows spatial harmonic emissions for two tightly focused beams at two pressures, and they are quite different for Type-1 Bessel and Gaussian beams.

A similar analysis for loosely focused laser beams is also carried out. For loose focusing Gaussian beam, the ionization level (in the end of the laser pulse) is found to be about 15 %, so the values of $\Delta k_{q,el} + \Delta k_{q,at}$ are about 1 and 3.5 mm^{-1} at 10 Torr for H15 and H35, respectively. These values increase to 8 and 28 mm^{-1} at 80 Torr. In Table 4.1, it does not change the phase mismatch much if the confocal parameter b changes to 15 mm for the Gaussian beam. $\Delta k_{q,geo} - K_{q,dip}$ is comparable with $\Delta k_{q,el} + \Delta k_{q,at}$ at 10 Torr, and $\Delta k_{q,el} + \Delta k_{q,at}$ becomes dominant as the pressure is increased. Figure 4.8 shows that even at moderate pressures spatial harmonic emissions for Type-2 Bessel and Gaussian beams are quite similar.

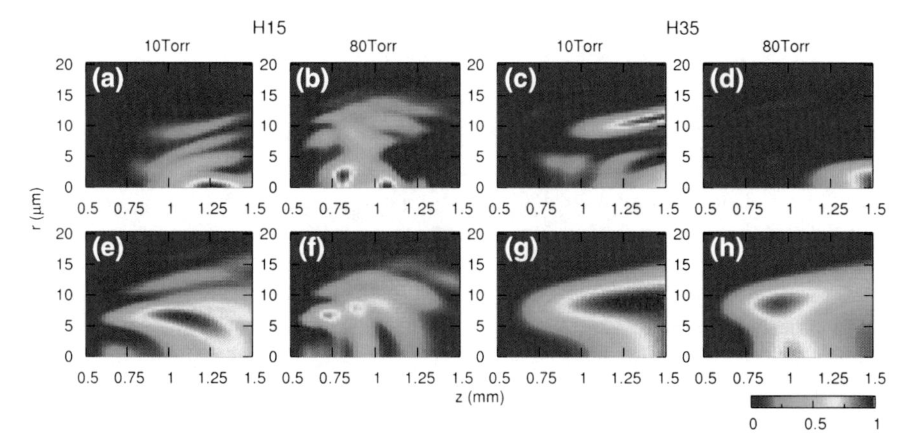

Fig. 4.7 Spatial distributions (normalized) of harmonic intensity under different pressures (10 and 80 Torr) using tight focusing laser beams. *Upper row*: Type-1 Bessel beam; *lower row*: Gaussian beam ($w_0 = 25\,\mu$m). The center of gas-jet (1-mm wide) is at $z_0 = 1$ mm. Adapted from [5]. © (2012) by the American Physical Society

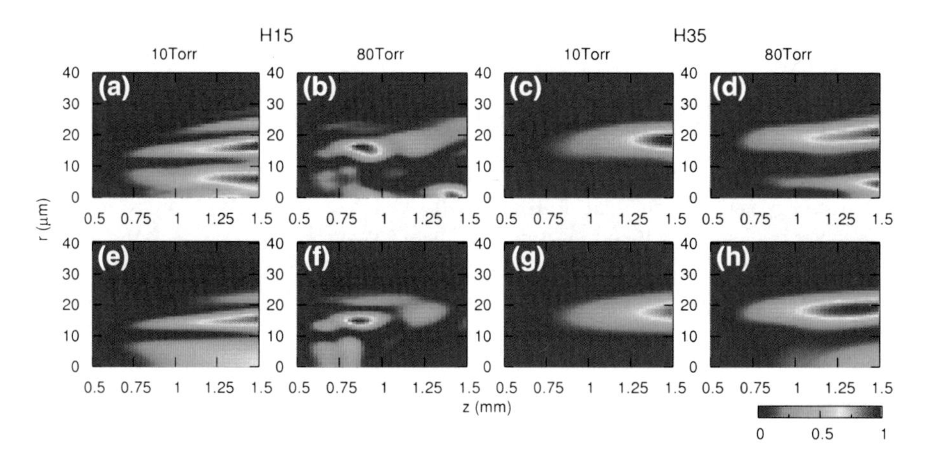

Fig. 4.8 Same as Fig. 4.7 except for loose focusing laser beams. *Upper row*: Type-2 Bessel beam; *lower row*: Gaussian beam ($w_0 = 50\,\mu$m). Adapted from [5]. © (2012) by the American Physical Society

4.5 Conclusion

In this chapter, I have examined the HHG for initial incident intense lasers (at the entrance of gas medium) that have Gaussian or truncated Bessel spatial profiles. With the inclusion of phase-matching and propagation effects, I have investigated how the harmonic emission depends on gas-jet position and gas pressure, for tightly and loosely focused Bessel and Gaussian beams. First, using 780-nm few-cycle pulses I have simulated HHG spectra of Ar reported in Wörner et al. [3]. No matter whether

I have assumed that the spatial profile was a truncated Bessel beam or a Gaussian beam, I was unable to reproduce the deep and broad Cooper minimum in the observed HHG spectrum of Ar. However, my simulation was able to reproduce the observed HHG spectrum of Ar in Shiner et al. [4] using 1800-nm lasers. Maybe additional experiments should be performed to clarify the existing discrepancy for 780-nm data.

Phase matching conditions for tightly and loosely focused Bessel and Gaussian beams have also been analyzed by varying gas-jet position and gas pressure. I was able to demonstrate that the harmonic growth maps were very similar thus resulting in nearly identical harmonic spectra for loosely focused Bessel or Gaussian beams. The situations were different for tightly focused beams. Harmonic growth maps were quite different for Bessel and Gaussian beams, so the resulting HHG spectra differed from each other as well. At higher pressure and/or intensity, the phase matching analysis is complicated due to laser defocusing and blue shift. To probe electronic structures of an atomic or molecular target by using the HHG as mentioned in Sect. 1.4.2, the harmonic spectra from loosely focused beams are preferable since they would be less sensitive to gas-jet location and other experimental parameters. For tightly focused beams, however, the harmonic spectra are very sensitive to experimental conditions. So the comparison of the theoretical simulation with the experiment is less straightforward since experimental parameters are not all well specified in general.

References

1. M. Nisoli, S. De Silvestri, O. Svelto, Generation of high energy 10 fs pulses by a new pulse compression technique. Appl. Phys. Lett. **68**, 2793–2795 (1996)
2. M. Nisoli, E. Priori, G. Sansone, S. Stagira, G. Cerullo, S. De Silvestri, C. Altucci, R. Bruzzese, C. de Lisio, P. Villoresi, L. Poletto, M. Pascolini, G. Tondello, High-brightness high-order harmonic generation by truncated bessel beams in the sub-10-fs regime. Phys. Rev. Lett. **88**, 033902 (2002)
3. H.J. Wörner, H. Niikura, J.B. Bertrand, P.B. Corkum, D.M. Villeneuve, Observation of electronic structure minima in high-harmonic generation. Phys. Rev. Lett. **102**, 103901 (2009)
4. A.D. Shiner, B.E. Schmidt, C. Trallero-Herrero, H.J. Wörner, S. Patchkovskii, P.B. Corkum, J.-C. Kieffer, F. Légaré, D.M. Villeneuve, Probing collective multi-electron dynamics in xenon with high-harmonic spectroscopy. Nature Phys. **7**, 464–467 (2011)
5. C. Jin, C.D. Lin, Comparison of high-order harmonic generation of Ar using truncated bessel and gaussian beams. Phys. Rev. A **85**, 033423 (2012)
6. J.A.R. Samson, W.C. Stolte, Precision measurements of the total photoionization cross-sections of He, Ne, Ar, Kr, and Xe. J. Electron Spectrosc. Relat. Phenom. **123**, 265–276 (2002)
7. S. Minemoto, T. Umegaki, Y. Oguchi, T. Morishita, A.T. Le, S. Watanabe, H. Sakai, Retrieving photorecombination cross sections of atoms from high-order harmonic spectra. Phys. Rev. A **78**, 061402 (2008)
8. P. Colosimo, G. Doumy, C.I. Blaga, J. Wheeler, C. Hauri, F. Catoire, J. Tate, R. Chirla, A.M. March, G.G. Paulus, H.G. Muller, P. Agostini, L.F. DiMauro, Scaling strong-field interactions towards the classical limit. Nature Phys. **4**, 386–389 (2008)
9. J.P. Farrell, L.S. Spector, B.K. McFarland, P.H. Bucksbaum, M. Gühr, M.B. Gaarde, K.J. Schafer, Influence of phase matching on the Cooper minimum in Ar high-order harmonic spectra. Phys. Rev. A **83**, 023420 (2011)

10. J. Higuet, H. Ruf, N. Thiré, R. Cireasa, E. Constant, E. Cormier, D. Descamps, E. Mével, S. Petit, B. Pons, Y. Mairesse, B. Fabre, High-order harmonic spectroscopy of the Cooper minimum in argon: Experimental and theoretical study. Phys. Rev. A **83**, 053401 (2011)
11. H.G. Muller, Numerical simulation of high-order above-threshold-ionization enhancement in argon. Phys. Rev. A **60**, 1341–1350 (1999)
12. C. Jin, H.J. Wörner, V. Tosa, A.T. Le, J.B. Bertrand, R.R. Lucchese, P.B. Corkum, D.M. Villeneuve, C.D. Lin, Separation of target structure and medium propagation effects in high-harmonic generation. J. Phys. B **44**, 095601 (2011)
13. Ph. Balcou, P. Salières, A. L'Huillier, M. Lewenstein, Generalized phase-matching conditions for high harmonics: the role of field-gradient forces. Phys. Rev. A **55**, 32043210 (1997)
14. H. Dachraoui, T. Auguste, A. Helmstedt, P. Bartz, M. Michelswirth, N. Mueller, W. Pfeiffer, P. Salières, U. Heinzmann, Interplay between absorption, dispersion and refraction in high-order harmonic generation. J. Phys. B **42**, 175402 (2009)
15. M.B. Gaarde, J.L. Tate, K.J. Schafer, Macroscopic aspects of attosecond pulse generation. J. Phys. B **41**, 132001 (2008)
16. T. Pfeifer, C. Spielmann, G. Gerber, Femtosecond x-ray science. Rep. Prog. Phys. **69**, 443–505 (2006)
17. C. Winterfeldt, C. Spielmann, G. Gerber, Colloquium: Optimal control of high-harmonic generation. Rev. Mod. Phys. **80**, 117–140 (2008)
18. L.E. Chipperfield, P.L. Knight, J.W.G. Tisch, J.P. Marangos, Tracking individual electron trajectories in a high harmonic spectrum. Opt. Commun. **264**, 494–501 (2006)
19. S. Kazamias, S. Daboussi, O. Guilbaud, K. Cassou, D. Ros, B. Cros, G. Maynard, Pressure-induced phase matching in high-order harmonic generation. Phys. Rev. A **83**, 063405 (2011)

Chapter 5
Generation of an Isolated Attosecond Pulse in the Far Field by Spatial Filtering with an Intense Few-Cycle Mid-infrared Laser

5.1 Introduction

As discussed in Sect. 1.4.1, high-order harmonic generation (HHG) has been widely used to produce attosecond pulses in the extreme ultraviolet (XUV) or soft X-ray [1–3]. Due to its great potential for probing ultrafast electronic processes, there is a plethora of techniques available to generate an isolated attosecond pulse (IAP), with the idea that high harmonics could be generated from half an optical cycle only in a few- or multi-cycle infrared laser pulse. Since the harmonic field generated by all atoms within the laser focus co-propagates with the fundamental laser pulse in the gas medium, as well as the possible further propagation in free space depending on the experimental setup, any methods using HHG to generate the IAP also need to take the effects of macroscopic propagation of fundamental and harmonic fields into account. As demonstrated in Chaps. 3 and 4, these effects have been well taken care of by using the well-established propagation theory with the quantitative rescattering (QRS)-based single-atom induced dipoles, the phase of high harmonic which is inevitably involved in the propagated harmonic field plays an essential role for the attosecond pulse generation. In this chapter, I will focus on the generation of attosecond pulses, which allows one to test the phase of high harmonic in previous studies as well.

Recently, Xe atom has become attractive for generating an intense IAP [4], probing the multi-electron dynamics with the high-harmonic spectroscopy [5], and studying the phase-matching effects in the generation of high-energy photons [6]. Using low-order harmonics of Xe by a CEP-stabilized laser, Ferrari et al. [4] reported that they were able to generate a high-energy 160-as IAP. Very dilute gas was used in their experiment so that the fundamental field was not severely distorted, but the ground state of atom was depleted very quickly in the leading edge of laser pulse due to the high laser intensity applied. Only low-order harmonics emitted within one half cycle were synthesized to obtain an IAP. Shiner et al. [5] generated HHG spectra of Xe up to the photon energy of 160 eV using a 1.8-μm laser with a duration of less than two optical cycles. They have shown that there was a strong enhancement above about 90 eV exhibited in the HHG spectrum. This enhancement has been well-known in

C. Jin, *Theory of Nonlinear Propagation of High Harmonics Generated in a Gaseous Medium*, Springer Theses, DOI: 10.1007/978-3-319-01625-2_5,

photoionization (PI) of Xe. The differential PI cross section of $5p$ shell of Xe is modified due to the presence of a strong shape resonance from $4d$ shell through the channel coupling—a feature attributed to the many-electron effects. According to the QRS theory, such enhancement is anticipated in the single-atom HHG spectrum since the differential photorecombination (PR) cross section (related to photoionization) enters directly in the laser-induced dipole. To simulate high harmonics at high-photon energies, multi-electron effects thus have to be incorporated into laser-induced dipoles. Using such induced dipoles in the QRS model, I simulate HHG spectra of Xe generated by 1.8-μm lasers with including the macroscopic propagation and phase-matching effects.

In this chapter, I mostly aim at understanding HHG spectra of Xe observed experimentally in [7], which show nearly continuous photon-energy distributions (to be called continuum structure) at high laser intensities. Such continuum spectra have also been observed in molecules, like NO [7]. From my simulation, I wish to investigate whether IAPs are generated by these harmonics. For this purpose, I demonstrate how to synthesize an IAP with different ranges of high harmonics by using a spatial filter in the far field. The approach proposed here is different from that in Ferrari et al. [4], but similar to the analysis in Gaarde et al. [8]. In Sect. 5.2, I will briefly summarize the QRS theory including the multi-electron effects and present calculated HHG spectra of Xe with a mid-infrared laser. In Sect. 5.3, I will show the spatiotemporal electric field of fundamental laser pulse. In Sect. 5.4, I will first give a wavelet theory for the time-frequency analysis, and then plot near- and far-field harmonics in time domain using this technique. In Sect. 5.5, I will demonstrate the IAP generation in the far field by synthesizing the harmonics from order 40 to 80 (H40–H80) and H90-H130. In Sect. 5.6, a study of carrier-envelope phase (CEP) dependence of the IAP will be presented and will conclude that it is still possible to obtain an IAP even by using a laser where the CEP is not stabilized. I will also compare the attosecond pulses calculated using QRS and SFA in Sect. 5.7. A short summary in Sect. 5.8 will conclude this chapter.

5.2 Macroscopic HHG Spectra of Xe Using an 1825-nm Few-Cycle Laser

5.2.1 Photorecombination Dipole Moment of Xe in the QRS Theory

Single-atom induced dipole moment $D(t)$ served to the macroscopic propagation in Eq. (2.53) is obtained by the QRS theory. In the energy (or frequency) domain, $D(\omega) = W(\omega)d(\omega)$, where $d(\omega)$ is the PR transition dipole moment and $W(\omega)$ is the microscopic electron wave packet. In the QRS theory, $W(\omega)$ is determined by the laser field solely and can be accurately calculated based on the strong-field approximation (SFA), and $d(\omega)$ is the transition dipole between initial and final states of PR or PI. The transition dipole $d(\omega)$ is usually calculated under single-active electron (SAE) approximation when the multi-electron effect is not important. However, as routinely done in the PI theory of atoms and molecules, the transition dipole is easily generalized to include the many-electron effects. There have been some well-established multi-channel calculations to include the many-electron effects in $d(\omega)$, such as many-body perturbation theory, close-coupling method, R-matrix method, random-phase approximation, and so on. Photoionization of Xe has been well studied, $d(\omega)$ of Xe in this chapter is obtained semi-empirically. The major many-electron effect of Xe in the PI from $5p$ shell occurs at the photon energy where $4d$ shell is open. Thus below about 60 eV, the transition dipole from $5p$ can be accurately obtained from a single-electron model without considering the many-electron effect. Note that both magnitude and phase are involved in the transition dipole. At higher energies, the PI cross section of Xe from $4d$ has a large and broad shape resonance around 100 eV, so the effect from $4d$ shell on the transition dipole of $5p$ becomes important. Consequently, the inter-shell coupling enhances $d(\omega)$ for $5p$ near and above 90 eV. Using the relativistic random-phase approximation (RRPA), Kutzner et al. [9] have calculated such enhancement. In my calculation, the magnitude of $d(\omega)$ is taken from [9] while the phase is taken from $5p$ shell under the SAE approximation. This will not change the temporal structure of attosecond pulses (will be shown in Sect. 5.7) even with the pseudo phase at high-photon energies since the phase of $D(\omega)$ is dominated by the phase of electron wave packet $W(\omega)$. Comment that the QRS theory deals with single-atom induced dipole in the energy domain, thus its calculation is similar to time-independent theory in the PI, which has been well-established in the last 30 years.

5.2.2 Macroscopic HHG Spectra of Xe at Low and High Intensities

Experimentally HHG spectra of Xe have been extended to the photon energy of over one hundred electron volts using 1.8-μm lasers with the pulse duration of few optical cycles recently [5, 7].

In Fig. 5.1, it shows the simulated HHG spectra of Xe exposed to a 14-fs (FWHM) and 1825-nm laser. The simulation parameters are chosen to be as close as those in the experiment of Trallero-Herrero et al. [7]. Laser beam waist is 100 μm. A 1-mm-long gas jet is placed at the laser focus. Gas pressure is 30 Torr. A slit with a width of 190 μm is placed 455 mm behind the laser focus. Only high harmonics after the slit are detected. For the present purpose, I only analyze the theoretical HHG spectra obtained at two laser peak intensities 0.5×10^{14} W/cm^2 and 1.0×10^{14} W/cm^2, which are below and above the critical intensity for Xe at $\sim0.87\times10^{14}$ W/cm^2 [11], respectively.[1] Here the critical field is defined with respect to the static electric field where an electron can classically escape over the top of the field-induced potential barrier, see Eq. (1.1).

[1] The detailed comparison between experiment and theory for many laser intensities is in [7].

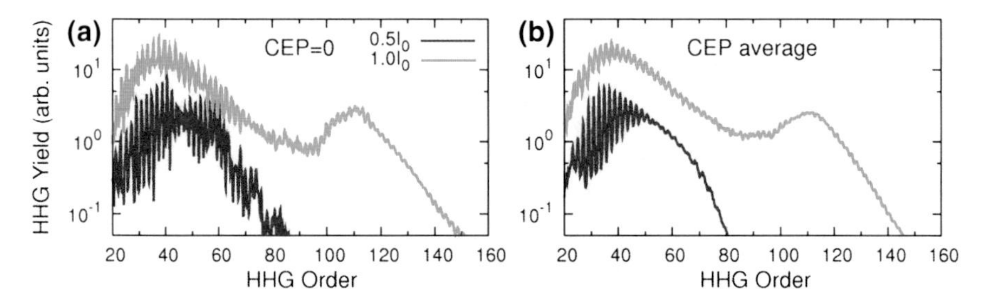

Fig. 5.1 Macroscopic HHG spectra of Xe in an 1825-nm laser, for **a** CEP=0 and **b** CEP averaged. Laser intensities are indicated in units of $I_0 = 10^{14}$ W/cm^2. See text for additional laser parameters and the experimental arrangement. Adapted from [10]. © (2011) by the American Physical Society

In Fig. 5.1a, CEP=0. High harmonics resulted from two laser intensities present different characteristics. At low intensity, the valley between neighboring odd harmonics is very deep, so the harmonics are very sharp. For high intensity, the spectrum shows a continuum structure, i.e., the valley is very shallow. Furthermore, due to the blue shift of fundamental field, the harmonics are not exactly at odd integers of the fundamental frequency. Note that inter-shell or many-electron effects discussed in Sect. 5.2.1 lead to the rise of spectrum above H90 (about 62 eV). HHG spectra have a strong CEP dependence due to a few-cycle laser pulse applied. In Fig. 5.1b, CEP averaged HHG spectra are shown. Main characteristics of high harmonics remain the same as CEP=0 except that the harmonic spectra are much smoother. In following sections the CEP is fixed at zero unless otherwise stated.

5.3 Spatiotemporal Evolution of Fundamental Laser Field

To understand different spectral features at low and high laser intensities in Fig. 5.1, I inspect the fundamental laser pulse in the ionizing medium. In Fig. 5.2, spatiotemporal intensity profiles and on-axis electric fields of laser pulse at the entrance and the exit of gas jet are shown. Laser peak intensity is chosen as 1.0×10^{14} W/cm^2. According to an empirical ADK formula in the barrier-suppression regime [11], this gives an ionization probability of ∼35 % at the end of laser pulse for Xe. While at the entrance of gas jet laser pulse has a good Gaussian shape both in time and space, it is strongly reshaped when it is propagated in the ionizing medium. At the exit it shows the positive chirp in time (blue shift in frequency), see Fig. 5.2c, and defocusing in space, see Fig. 5.2b. The fundamental laser field with peak intensity of 0.5×10^{14} W/cm^2 is also checked. It always maintains the Gaussian spatial distribution without any blue shift through the propagation in the medium because the ionization probability is too low. It concludes that the spatiotemporal reshaping of fundamental laser field at high intensity is responsible for the continuum structure

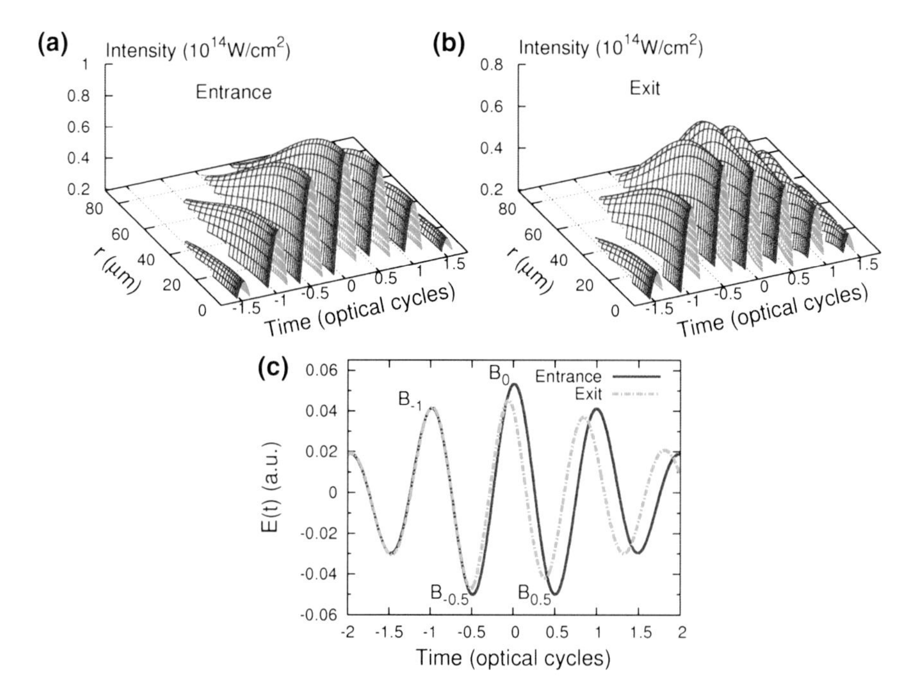

Fig. 5.2 Spatiotemporal intensity profile of fundamental laser pulse at **a** the entrance and **b** the exit of a Xe gas jet. **c** Evolution of on-axis electric field at the entrance (*solid line*) and the exit (*dot-dashed line*). Laser field becomes chirped during the propagation. For sub-cycle dynamics analysis, labels B_t, with $t = -1, -0.5, 0$ and 0.5 (in units of optical cycles) are used to indicate the approximate half-cycle where an electron is born. Note that t is defined within the half cycle only. Laser intensity at the focus is 1.0×10^{14} W/cm^2 (assumed in the vacuum) and CEP=0 in these figures. Adapted from [10]. © (2011) by the American Physical Society

in HHG spectra in Fig. 5.1. Note that Gaarde et al. [12] have obtained similar results by using a 750-nm laser interacting with Ne gas.

5.4 Time-Frequency Representation of High Harmonics

5.4.1 Wavelet Analysis of Attosecond Pulses

Harmonic field $E_{\mathrm{h}}(t)$ can be presented in a time-frequency representation (TFR) (or spectrogram), which is a simultaneous representation of temporal and spectral characteristics of high harmonics. I perform the time-frequency analysis by using the wavelet transform of harmonic field [13–16]:

$$A(t, \omega) = \int E_{\mathrm{h}}(t') w_{t,\omega}(t') dt'. \tag{5.1}$$

Here the wavelet kernel is $w_{t,\omega}(t') = \sqrt{\omega}W[\omega(t'-t)]$, and the Morlet wavelet is chosen as [13]:

$$W(x) = (1/\sqrt{\tau})e^{ix}e^{-x^2/2\tau^2}. \tag{5.2}$$

The width of window function (or the wavelet kernel) in the wavelet transform varies with frequency, but the number of oscillations within the window, (which is proportional to τ), is held a constant if τ is fixed. So the absolute value of $A(t, \omega)$ depends on τ. The dependence of $A(t, \omega)$ on parameter τ has been tested, the general temporal pattern didn't change much with τ. In this chapter, $\tau = 15$ is chosen to perform the wavelet analysis.

Near-field harmonics emitted at the exit face of medium can be considered as a source for far-field harmonics, which also have the spatial dependence along radial direction. Harmonic spatial distribution in the near field could be complicated, see Fig. 4 in [17]. In order to avoid this complexity, I first calculate near-field $A(t, \omega)$ for each radial point, and then do the integral over radial coordinate [15]:

$$|A_{\mathrm{near}}(t, \omega)|^2 = \int_0^{\infty} 2\pi r dr \left| \int E_{\mathrm{h}}(r, t')w_{t,\omega}(t')dt' \right|^2. \tag{5.3}$$

To illustrate the divergence of high harmonics, I preform the TFR for far-field harmonics at each radial point only (in a plane perpendicular to the propagation axis, which is far away from the medium).

Here are two filters applied in this chapter. The first one is the spectral filter used to select a range of high harmonics ($\omega_1 - \omega_2$) to produce attosecond pulse trains (APTs) or IAPs. Theoretically one can obtain the total intensity of an APT or an IAP in the near field (after radial integration) as following [18]:

$$I_{\mathrm{near}}(t) = \int_0^{\infty} 2\pi r dr \left| \int_{\omega_1}^{\omega_2} E_{\mathrm{h}}(r, \omega)e^{i\omega t}d\omega \right|^2. \tag{5.4}$$

In the far field, the second filter is used to select high harmonics in a prescribed area, which is a spatial filter. In this chapter, I assume that this filter is a circular one with a radius of r_0, and is placed perpendicular to the harmonic propagation direction. The intensity of an APT or an IAP in the far field is expressed as

$$I_{\mathrm{far}}(t) = \int_0^{r_0} 2\pi r dr \left| \int_{\omega_1}^{\omega_2} E_{\mathrm{h}}^{\mathrm{f}}(r, \omega)e^{i\omega t}d\omega \right|^2. \tag{5.5}$$

5.4.2 Time-Frequency Analysis of High Harmonics in Near and Far Fields

As shown in Eq. (1.8), for each harmonic order q, the induced-dipole phase can be expressed as $\varphi_i^q(r, z, t) = -\alpha_i^q I(r, z, t)$. Here $I(r, z, t)$ is the spatiotemporal intensity of fundamental laser field, and the proportional constant $\alpha_{i=S,\ L}$ depends on electron following the "short" (S) or "long" (L) trajectory in laser field. This phase can also be related to ponderomotive energy U_p and electron excursion time τ_i^q by $\varphi_i^q \approx -\beta_i U_p \tau_i^q$ [19], where the coefficient β_i is much smaller for the "short" trajectory than that for the "long" trajectory. Roughly speaking, the electron excursion times for two trajectories are $\tau_S^q \approx T/2$ and $\tau_L^q \approx T$ (T is the laser period) [20]. One can see that the phase (both "short" and "long" trajectories) grows with the cubic power of wavelength in the first-order approximation. The radial variation $\partial\varphi_i^q(r)/\partial r$ causes the curvature of phase front, and makes the harmonic beam divergent. The divergence of "short"- or "long"-trajectory harmonic is determined by either $\partial\alpha_i^q/\partial r$ or $\partial I(r)/\partial r$.

1. High harmonics in the near field

Time-frequency representations, $|A_{\mathrm{near}}(t, \omega)|^2$, defined in Eq. (5.3) are shown in Fig. 5.3a and d. High harmonics above H40 at two laser intensities are collected at the exit face of gas jet (near field). In Fig. 5.3a, the first (earliest) group of generated harmonics are indicated by symbols S and L. Here S (L) stands for "short" ("long")-trajectory harmonics that have positive (negative) chirp. In this chapter the time is always defined in the moving coordinate frame, also see [17]. B_{-1} is used to indicate high harmonics from electrons born at $t = -1$ (in units of optical cycles) in the leading edge of laser pulse, see Fig. 5.2c.[2] In the following, B_t indicates the electron born time t (in units of optical cycles) in the figure, while the harmonic emission time (or the electron recombination time) is read off from the horizontal axis of figure, one for the "short", and the other for the "long" trajectory. At low intensity in Fig. 5.3a, one can see that for electrons born at $t = -1, -0.5, 0$ and 0.5, both S and L contribute to the generated harmonics, i.e., harmonics are generated by electrons born over four half cycles. Note that Tate et al. [21] have shown that for high harmonics generated by mid-infrared lasers, there were unexpected contributions from electron trajectories even longer than "long" trajectories in the single-atom response. But none of these higher-order trajectories is presented in the figure. This is probably because these trajectories are all eliminated during the propagation in the medium due to their large phases. At low intensity, contributions from "long" trajectories cannot be eliminated even after the propagation in the medium.

In Fig. 5.3d, the similar TFR analysis for high intensity is carried out. Since laser intensity is twice higher than that in Fig. 5.3a, higher harmonic cutoff from

[2] Electrons are mostly born off the sub-cycle peaks as shown in Fig. 1.5, so $t = -1$ only roughly tells electrons' birth time.

each burst is easily seen. It is distinct that for electrons born at $t = -1$ and -0.5, i.e., from the leading edge of pulse, there are no contributions to the harmonics from "long" trajectories. Because the laser intensity is twice higher, the phase of each harmonic is also twice higher, see Fig. 17 in [22] and Fig. 1A in [23]. This would result in cancellation of contributions from "long" trajectories. For electrons born at the falling edge of laser pulse, the phases of high harmonics due to "long" trajectories are smaller because of blue shift (thus shorter wavelength) and reshaping (thus lower intensity), so these "long" trajectories can survive after the propagation in the medium, for example, for electrons born at $t = 0$ and 0.5.

2. On-axis harmonics in the far field

TFRs are shown in Fig. 5.3b and e for $r = 0$ mm in the far field (455 mm after laser focus). At low intensity, the emissions from "short" trajectories at different times have the similar small divergence, and they all survive along the axis in the far field after the propagation in free space. The interference between "short"-trajectory harmonics from each half cycle could lead to enhancement in odd harmonics and suppression in even harmonics.[3] So it can explain the big contrast between an odd harmonic and neighboring harmonics shown in Fig. 5.1a for HHG spectra obtained with a slit. At high intensity, only "short"-trajectory harmonics born at $t = -1$ survive (the next one at $t = -0.5$ is much weaker) on axis after a long-distance free propagation. The harmonic emission is nearly limited to half an optical cycle, and this would result in a nearly continuum spectrum, probably can be used for generating an isolated attosecond pulse.

3. Off-axis harmonics in the far field

TFRs are shown for $r = 1$ mm (divergence: 2.2 mrad) in the far field in Fig. 5.3c and f. These plots show an obvious time delay for each off-axis burst with respect to the on-axis burst because off-axis harmonic travels a longer distance in free space. At low intensity, high harmonics from "long" trajectories are presented on each burst because they have large divergence. At high intensity, "short" trajectories contribute to bursts $B_{-0.5}$ and B_0, originating from the laser pulse reshaping, see Fig. 5.2b showing laser peak intensity shifting to the region away from the propagation axis. They are expected to experience larger $\Delta I(r)$ (larger divergence) with respect to "short"-trajectory electrons born at B_{-1} at the leading edge. One can also expect that the continuum harmonic spectrum from "short" trajectories for electrons born at $t = -0.5$ is generated off axis in Fig. 5.3f.

Note that attochirp [23, 25] (i.e., the harmonic emission time varying with harmonic order) of "short"- or "long"-trajectory harmonics still exists even after the propagation in the gas medium or in free space. The positive attochirp may be

[3] This mechanism has been illustrated in Fig. 18 of [24].

compensated using a "plasma compressor" [23] because the free electrons induce a negative group velocity dispersion, or by thin filters with the linear negative group velocity dispersion [26]. The attochirp in the single-atom level is inversely proportional to laser wavelength [27]. This implies that using a long-wavelength laser, for example, an 1825-nm laser, one can select a broad range of high harmonics to synthesize a short attosecond pulse. In Fig. 5.3e and f, the harmonic emission of "short" trajectory in the far field varies with time or radial distance. This provides possibilities to apply spectral and spatial filters to generate IAPs using different ranges of high harmonics on or off axis. I will only show that these two filters applied on axis in the far field in the following.

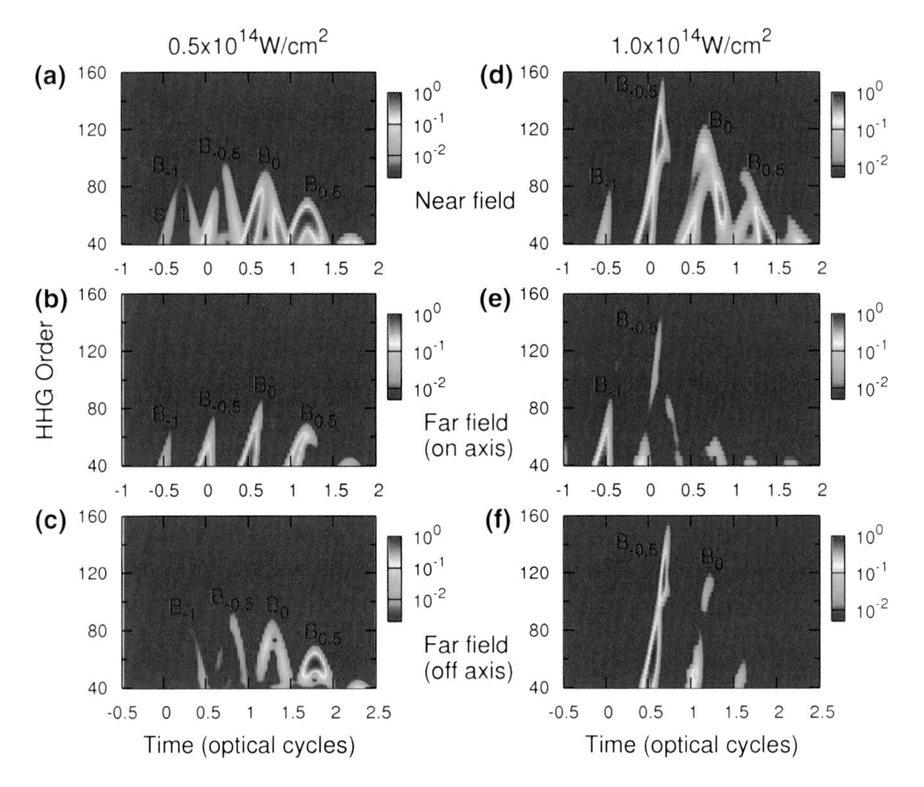

Fig. 5.3 *Top row*: Time-frequency representation (TFR) of high harmonics in the near field. *Middle row*: TFR for on-axis ($r = 0$ mm, divergence: 0 mrad) harmonics in the far field. *Bottom row*: TFR for off-axis ($r = 1$ mm, divergence: 2.2 mrad) harmonics in the far field. Far-field position is at $z = 455$ mm, laser intensity along each column is indicated, and CEP = 0. Electrons are released at each half cycle, labeled by B_t, with $t = -1, -0.5, 0$ and 0.5 as in Fig. 5.2. For each B_t, electrons can follow either a "short" (S) or a "long" (L) trajectory to recombine with the ion. Harmonic emission time can be read from time axis. For each B_t, the emission time of off-axis harmonic is later than that of on-axis harmonic, e.g., compare **b** versus **c** and **e** versus **f**. All TFRs have been normalized. Adapted from [10]. © (2011) by the American Physical Society

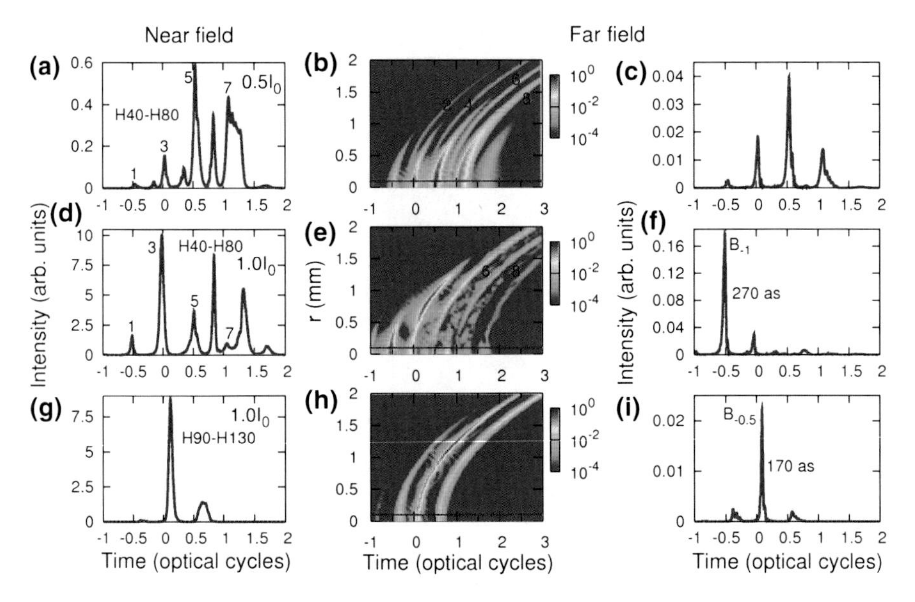

Fig. 5.4 *First column*: Intensity (or envelope) of attosecond pulses in the near field, synthesized from high harmonics and laser intensity shown in each frame. Laser intensities are given in units of $I_0 = 10^{14}$ W/cm^2. In **a** and **d**, odd bursts ("short" trajectories) are labeled. Even bursts due to "long" trajectories are not labeled for brevity. *Middle column*: Spatial distribution (normalized) of attosecond pulses in the far field ($z = 455$ mm). Note that even bursts ("long" trajectories) have large divergence, or at large r. Odd bursts (not labeled) have smaller divergence. There is a time delay between off-axis attosecond pulses compared to on-axis ones. *Last column*: Intensity of attosecond pulses in the far field using a spatial filter with a radius $r_0 = 100\,\mu$m (indicated by the *solid line* in *red* in each *middle-column* frame). Note that high harmonics and laser intensity are the same in the same row. Adapted from [10]. © (2011) by the American Physical Society

5.5 Spectral and Spatial Filtering in the Generation of Attosecond Pulses

To synthesize attosecond pulses, the spectral filtering is one of generally used methods. In this section I also study how attosecond pulses are manipulated by another method, i.e., the spatial filtering. In Fig. 5.4a, it displays the intensity profile of an XUV light by synthesizing H40–H80 at the near field generated at low laser intensity of 0.5×10^{14} W/cm^2. Using Eq. (5.4), the intensity of attosecond pulses $I_{near}(t)$ can be obtained. The time-frequency analysis of these harmonics has been given in Fig. 5.3a. One can observe that attosecond bursts occur at each half optical cycle, which can be attributed to high harmonics resulting from "short" trajectories, and other pulses present in between which are attributed to contributions from "long" trajectories. In the figure main peaks from "short" trajectories are labeled by 1, 3, 5 and 7, while those from "long" trajectories (2, 4, 6 and 8) in between are not labeled. Generated attosecond pulses thus show a poor periodicity in time, see Fig. 5.4a.

If attosecond pulses are synthesized in the far field, it may be possible to remove high harmonics resulting from "long" trajectories by introducing a spatial filter. In Fig. 5.4b, it shows intensity distributions of the synthesized light in space in the far field. Peaks 2, 4, 6 and 8 indicated in Fig. 5.4b, not shown in Fig. 5.4a explicitly, are attributed to "long" trajectories, so they are distributed far from the propagation axis. By using a spatial filter (indicated by a solid line in red, with a radius $r_0 = 100\,\mu\text{m}$) to select harmonics generated near the axis only, well-behaved APTs are then obtained as shown in Fig. 5.4c by using Eq. (5.5) to calculate $I_{far}(t)$. Note that the curved spatial distribution in Fig. 5.4b is caused by time delay between off-axis and on-axis harmonics, and it can be understood mathematically since each harmonic behaves like a Gaussian beam, and the geometric phase of each harmonic is proportional to r^2 along the transverse direction (or radial direction), see Fig. 4 in [17]. The extra traveling distance of off-axis harmonics can be compensated by a detector with a curved surface or by using a reflecting mirror to refocus the harmonic beam. In principle, if a spatial filter with a large radius is applied, this compensation becomes important to reduce the duration of attosecond pulses. In this chapter, to avoid this curvature effect the radius of spatial filter is chosen to be small enough.

Next the same range of high harmonics (H40–H80) generated at high laser intensity of 1.0×10^{14} W/cm^2 is used to synthesize attosecond pulses in the near field. As shown in Fig. 5.3d, "short" trajectories and "long" trajectories dominate the harmonic generation in the leading edge and the falling edge of laser pulse, respectively. The synthesized XUV light, shown in Fig. 5.4d indeed reflects this point where the first two peaks (1 and 3) occur at multiples of half optical cycles, while two strong peaks (not 5 and 7) among last four peaks are not. In Fig. 5.4e, the spatial distribution of synthesized XUV light in the far field indeed supports this description, and 6 and 8 indicate the "long"-trajectory emission. By using a spatial filter (indicated by a solid line in red, with a radius r_0=100 μm), only "short" trajectories are selected. As shown in Fig. 5.4f, a nice IAP with a duration of 270 as is obtained, accompanied by a sub peak with a much weaker intensity. This demonstrates the generation of IAPs using the spatial filtering. Strelkov et al. [28, 29] have proposed a similar mechanism of the IAP generation by using high harmonics in the plateau region generated by Ar gas at very high pressure.

TFR in Fig. 5.3e shows that at burst $B_{-0.5}$ there is the considerable on-axis emission above H80. In Fig. 5.4g I show the generated attosecond pulses by synthesizing H90-H130 in the near field. Both bursts having considerable contributions from "short" trajectories are shifted in time compared to 3 and 5 in Fig. 5.4d due to the positive attochirp. In the far field, see Fig. 5.4h, different divergences are shown for three bursts as discussed before. Finally, an IAP is obtained with a duration of about 170 as in Fig. 5.4i with a spatial filter. However, the intensity of IAP (with the higher central frequency) is about 1/8 as that in Fig. 5.4f due to, not only the larger divergence of "short"-trajectory harmonics born at $B_{-0.5}$ than at B_{-1}, but also the lower harmonic intensity of H90-H130 than that of H40–H80. On the other hand, the duration of IAP is decreased by 100 as. Gaarde et al. [8, 12] have proposed the similar mechanism of IAP generation by using high harmonics in the cutoff region by a 750-nm laser exposed on Ne gas.

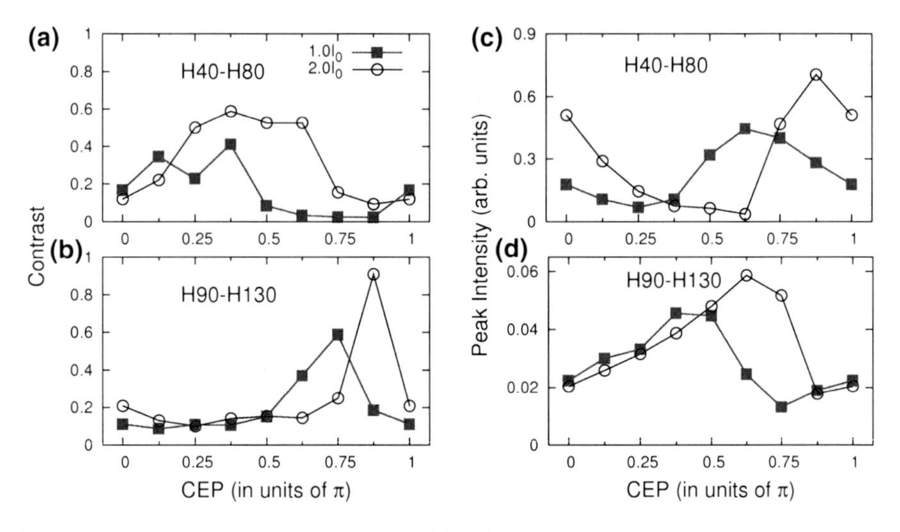

Fig. 5.5 **a** and **b**: Contrast ratio between intensities of the strongest satellite and the main attosecond burst as CEP is varied, **c** and **d**: Peak intensity of the main attosecond burst as a function of CEP. Harmonics used to generate an IAP are labeled. Laser intensities are shown in units of $I_0 = 10^{14}$ W/cm^2. Far-field position: $z = 455$ mm, and the radius of spatial filter: $r_0 = 100\,\mu$m. Adapted from [10]. © (2011) by the American Physical Society

5.6 CEP Dependence of Isolated Attosecond Pulses

The above discussed IAP generation in the far field by the spatial filtering is only for a single CEP, and thus it only can be realized in the experiment if the laser is CEP-stabilized (unfortunately has not been achieved for 1.8-μm lasers yet). To further check if the spatial filtering method can be applied for lasers that are not CEP-stabilized, I investigate the CEP dependence of IAP generation in this section.

I first show the contrast ratio between intensities of the strongest satellite and the strongest attosecond burst as a function of CEP in Fig. 5.5a and b, and then I show the peak intensity of the strongest attosecond burst in Fig. 5.5c and d as CEP is varied. Two laser intensities used in the analysis are indicated in the figure. A good IAP is to have the strong main peak and weak satellites. From Fig. 5.5c and d, at CEP's where the strongest attosecond bursts have high peak values, the contrast ratios shown in Fig. 5.5a and b are always small. In the meanwhile, the strongest attosecond burst is always weak when the contrast ratio is large. Based on the analysis in Fig. 5.5 with two laser intensities, one can conclude that it is still possible to generate single attosecond pulses even if the CEP of driving laser is not stabilized. This also explains the success in the generation of first single attosecond pulse where few-cycle laser pulses that were not phase-stabilized were applied [30].

5.7 Comparison Between QRS and SFA in Modeling Propagation Effects

In last two decades, the strong-field approximation (SFA), developed under the SAE approximation, has been widely used to predict the temporal structure of attosecond pulses even though it is well-known that SFA is unable to explain observed harmonic spectra precisely in general. In the present calculation, induced dipoles obtained by the QRS theory are inserted into propagation equations. For single-atom response, the QRS theory has been tested against the TDSE, both for magnitude and phase of harmonics, as documented in Le et al. [31], for example. In the QRS, wave packet (including both magnitude and phase) is obtained from the SFA. The free-field transition dipole $d(\omega)$ also introduces a phase. In the SFA, this phase is a constant, either real or pure imaginary (depending on the symmetry of ground state) and independent of the harmonic order. In the QRS, $d(\omega)$ is in general a complex number. It is known from the PI theory that the phase of transition dipole does not change much with the photon energy. Thus phases of high harmonics calculated from the QRS and the SFA do not differ significantly. This argument is based on single-atom theories, but it is also true for the macroscopic harmonics, see Fig. 5 in [17]. Since the harmonic phase (more precisely, the phase difference between consecutive harmonics) is much more important in synthesizing attosecond pulses [32], this explains why the propagation theory based on the SFA has been so successful in explaining the generation of attosecond pulses, in spite of its failure in predicting observed harmonic

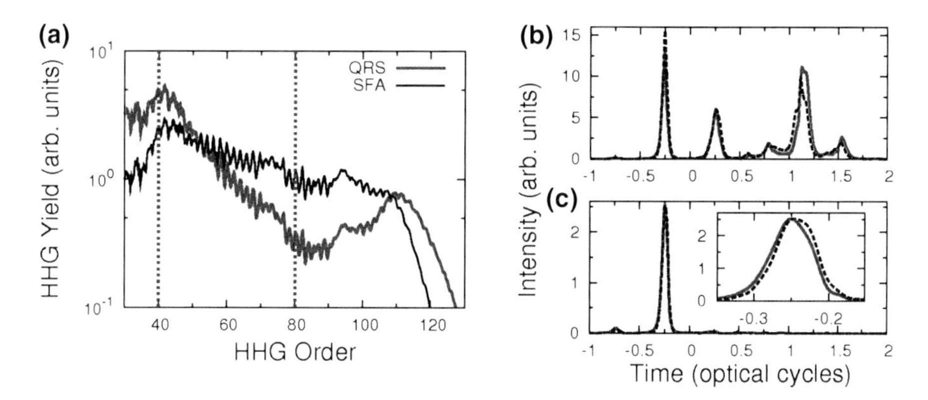

Fig. 5.6 Comparison of HHG spectra and attosecond pulses calculated using single-atom induced dipoles from the QRS and the SFA. **a** Macroscopic HHG spectra (total spectra without using a slit) of Xe by the QRS (*red line*) and the SFA (*black line*). Intensity of attosecond pulses **b** in the near field, and **c** in the far field ($z = 455$ mm) using a spatial filter with a radius $r_0 = 300\,\mu$m: QRS (*red-solid lines*) versus SFA (*black-dashed lines*). Inset in **c**: Enlarged temporal structure of an IAP. The spectra of two models are normalized at the peaks in *c*, this normalization factor is also used in *a* and *b*. Laser parameters: $I = 1.0 \times 10^{14}$ W/cm^2, CEP=$\pi/2$, and H40–H80 are used to synthesize the attosecond pulses. Adapted from [10]. © (2011) by the American Physical Society

spectra. In this section, I support this argument with actual results from the numerical simulations.

HHG spectra of Xe obtained from the SFA (within the SAE approximation) and the QRS (including multi-electron effects) are shown in Fig. 5.6a. Harmonic distributions with the harmonic order from two models differ greatly. Synthesized (H40–H80) attosecond pulses at the near field and far field are shown in Fig. 5.6b and c. Clearly attosecond pulses from two calculations are essentially identical (after an overall normalization), in spite of large differences in harmonic spectra. I have checked some other cases (different intensities and CEPs) and found that the temporal structures of attosecond pulses from two theories were always very similar. It is expected to see larger differences than those shown in Fig. 5.6b and c if a wider range of high harmonics are used or if the spectra from two theories differ much more, but the general conclusion is correct.

5.8 Conclusion

In this chapter, I have investigated the generation of isolated attosecond pulses (IAPs) in the far field by spatial filtering using a few-cycle mid-infrared laser at high intensities near and above the critical intensity of Xe. In the calculation of single-atom induced dipole moment, many-electron effects are included based on the QRS theory, specifically, the coupling of inner $4d$ shell of Xe in the differential $5p$ photorecombination transition dipole matrix element is included. By solving the Maxwell's wave equations, the nonlinear effects (such as dispersion, plasma effect and Kerr nonlinearity) of medium on the fundamental laser field is taken into account. These effects lead to the modification (or reshaping in space and time) of fundamental laser pulse. The spatiotemporal evolution of fundamental laser field in detail has been checked, and I have found that fundamental laser's reshaping was responsible for the continuum structure in the HHG spectrum at relatively high intensity. This conclusion was supported by the time-frequency analysis of high harmonics in both near and far fields.

Due to blue shift and defocusing of the fundamental laser pulse (or complicated reshaping), the divergence of harmonic emission from different half cycles is varied, I have shown that isolated attosecond pulses can be generated by synthesizing H40–H80 or H90-H130 with a spectral filter, selected by a spatial filter centered on the propagation axis in the far field. This approach works for a loosely focused laser at relatively high laser intensity (above the critical intensity), which is reshaped as it propagates through the medium with a moderate gas pressure. Since the ionization probability is required to be high enough, the mechanism of IAP generation in this chapter could be called as "ionization gating". Using a 750-nm laser interacting with 135-Torr Ne gas, a similar mechanism has been discussed by Gaarde et al. [8, 12]. However, the studies in this chapter have shown that using a long-wavelength laser with a moderate gas pressure ($\sim$ 30 Torr), it could easily reshape the fundamental laser field. On the other hand, the extended harmonic cutoff of Xe due to long-wavelength laser leads to a broad range of high harmonics available for the IAP generation. This approach is also different from Ferrari et al. [4] where only low harmonics ($\sim$ 30 eV, which is equivalent to H40 in this chapter) are used to generate the IAP. In addition, I have shown that the method proposed in this chapter is very robust and an IAP can be generated even with a CEP-not-stabilized laser.

References

1. F. Krausz, M. Ivanov, Attosecond physics. Rev. Mod. Phys. **81**, 163–234 (2009)
2. M.B. Gaarde, J.L. Tate, K.J. Schafer, Macroscopic aspects of attosecond pulse generation. J. Phys. B **41**, 132001 (2008)
3. T. Popmintchev, M.-C. Chen, P. Arpin, M.M. Murnane, H.C. Kapteyn, The attosecond nonlinear optics of bright coherent X-ray generation. Nat. Photon. **4**, 822–832 (2010)
4. F. Ferrari, F. Calegari, M. Lucchini, C. Vozzi, S. Stagira, G. Sansone, M. Nisoli, High-energy isolated attosecond pulses generated by above-saturation few-cycle fields. Nat. Photon. **4**, 875–879 (2010)
5. A.D. Shiner, B.E. Schmidt, C. Trallero-Herrero, H.J. Wörner, S. Patchkovskii, P.B. Corkum, J.-C. Kieffer, F. Légaré, D.M. Villeneuve, Probing collective multi-electron dynamics in xenon with high-harmonic spectroscopy. Nat. Phys. **7**, 464–467 (2011)
6. C. Vozzi, M. Negro, F. Calegari, S. Stagira, K. Kovács, V. Tosa, Phase-matching effects in the generation of high-energy photons by mid-infrared few-cycle laser pulses. New J. Phys. **13**, 073003 (2011)
7. C. Trallero-Herrero, C. Jin, B.E. Schmidt, A.D. Shiner, J.-C. Kieffer, P.B. Corkum, D.M. Villeneuve, C.D. Lin, F. Légaré, A.T. Le, Generation of broad XUV continuous high harmonic spectra and isolated attosecond pulses with intense mid-infrared lasers. J. Phys. B **45**, 011001 (2012)
8. M.B. Gaarde, K.J. Schafer, Generating single attosecond pulses via spatial filtering. Opt. Lett. **31**, 3188–3190 (2006)
9. M. Kutzner, V. Radojević, H.P. Kelly, Extended photoionization calculations for xenon. Phys. Rev. A **40**, 5052–5057 (1989)
10. C. Jin, A.T. Le, C.A. Trallero-Herrero, C.D. Lin, Generation of isolated attosecond pulses in the far field by spatial filtering with an intense few-cycle mid-infrared laser. Phys. Rev. A **84**, 043411 (2011)
11. X.M. Tong, C.D. Lin, Empirical formula for static field ionization rates of atoms and molecules by lasers in the barrier-suppression regime. J. Phys. B **38**, 2593–2600 (2005)
12. M.B. Gaarde, M. Murakami, R. Kienberger, Spatial separation of large dynamical blueshift and harmonic generation. Phys. Rev. A **74**, 053401 (2006)
13. X.M. Tong, S.-I. Chu, Probing the spectral and temporal structures of high-order harmonic generation in intense laser pulses. Phys. Rev. A **61**, 021802 (2000)
14. M.B. Gaarde, Ph Antoine, A. L'Huillier, K.J. Schafer, K.C. Kulander, Macroscopic studies of short-pulse high-order harmonic generation using the time-dependent Schrödinger equation. Phys. Rev. A **57**, 4553–4560 (1998)
15. M.B. Gaarde, Time-frequency representations of high order harmonics. Opt. Express **8**, 529–536 (2001)
16. V.S. Yakovlev, A. Scrinzi, High harmonic imaging of few-cycle laser pulses. Phys. Rev. Lett. **91**, 153901 (2003)
17. C. Jin, A.T. Le, C.D. Lin, Medium propagation effects in high-order harmonic generation of Ar and N_2. Phys. Rev. A **83**, 023411 (2011)
18. P. Antoine, A. L'Huillier, M. Lewenstein, P. Salières, B. Carré, Theory of high-order harmonic generation by an elliptically polarized laser field. Phys. Rev. A **53**, 1725–1745 (1996)

19. A. Zaïr, M. Holler, A. Guandalini, F. Schapper, J. Biegert, L. Gallmann, U. Keller, A.S. Wyatt, A. Monmayrant, I.A. Walmsley, E. Cormier, T. Auguste, J.P. Caumes, P. Salières, Quantum path interferences in high-order harmonic generation. Phys. Rev. Lett. **100**, 143902 (2008)
20. M. Lewenstein, P. Salières, and A. L'Huillier, Phase of the atomic polarization in high-order harmonic generation. Phys. Rev. A **52**, 4747–4754 (1995)
21. J. Tate, T. Auguste, H.G. Muller, P. Salières, P. Agostini, L.F. DiMauro, Scaling of wave-packet dynamics in an intense midinfrared field. Phys. Rev. Lett. **98**, 013901 (2007)
22. G. Sansone, C. Vozzi, S. Stagira, M. Nisoli, Nonadiabatic quantum path analysis of high-order harmonic generation: role of the carrier-envelope phase on short and long paths. Phys. Rev. A **70**, 013411 (2004)
23. Y. Mairesse, A. de Bohan, L.J. Frasinski, H. Merdji, L.C. Dinu, P. Monchicourt, P. Breger, M. Kovačev, R. Taieb, B. Carre, H.G. Muller, P. Agostini, P. Salières, Attosecond synchronization of high-harmonic soft X-rays. Science **302**, 1540–1543 (2003)
24. M. Protopapas, C.H. Keitel, P.L. Knight, Atomic physics with super-high intensity lasers. Rep. Prog. Phys. **60**, 389–486 (1997)
25. Y. Mairesse, A. de Bohan, L.J. Frasinski, H. Merdji, L.C. Dinu, P. Monchicourt, P. Breger, M. Kovacev, T. Auguste, B. Carré, H.G. Muller, P. Agostini, P. Salières, Optimization of attosecond pulse generation. Phys. Rev. Lett. **93**, 163901 (2004)
26. R. López-Martens, K. Varjú, P. Johnsson, J. Mauritsson, Y. Mairesse, P. Salières, M.B. Gaarde, K.J. Schafer, A. Persson, S. Svanberg, C.-G. Wahlström, A. L'Huillier, Amplitude and phase control of attosecond light pulses. Phys. Rev. Lett. **94**, 033001 (2005)
27. G. Doumy, J. Wheeler, C. Roedig, R. Chirla, P. Agostini, L.F. DiMauro, Attosecond synchronization of high-order harmonics from midinfrared drivers. Phys. Rev. Lett. **102**, 093002 (2009)
28. V. Strelkov, E. Mével, E. Constant, Isolated attosecond pulse generated by spatial shaping of femtosecond laser beam. Eur. Phys. J. Spec. Top. **175**, 15–20 (2009)
29. V.V. Strelkov, E. Mével, E. Constant, Generation of isolated attosecond pulses by spatial shaping of a femtosecond laser beam. New J. Phys. **10**, 083040 (2008)
30. M. Hentschel, R. Kienberger, Ch. Spielmann, G.A. Reider, N. Milosevic, T. Brabec, P. Corkum, U. Heinzmann, M. Drescher, F. Krausz, Attosecond metrology. Nature **414**, 509–513 (2001)
31. A.T. Le, T. Morishita, C.D. Lin, Extraction of the species-dependent dipole amplitude and phase from high-order harmonic spectra in rare-gas atoms. Phys. Rev. A **78**, 023814 (2008)
32. M.B. Gaarde, K.J. Schafer, Quantum path distributions for high-order harmonics in rare gas atoms. Phys. Rev. A **65**, 031406 (2002)

Chapter 6
Effects of Macroscopic Propagation and Multiple Molecular Orbitals on the High-Order Harmonic Generation of Aligned N_2 and CO_2 Molecules

6.1 Introduction

As discussed in Sect. 1.4.2, high-order harmonic generation (HHG) has been employed to probe the electronic structure of molecules on an ultrafast time scale [1–5]. The last step in the harmonic generation as shown in Sect. 1.2.1—"recombination" is the inverse process of photoionization. Any spectral features in the photoionization cross section (PICS) would thus be embodied in the high-harmonic spectrum as well. Nowadays molecules can be impulsively aligned by a femtosecond laser pulse [6], so one can further probe PICSs from aligned molecules by observing the HHG from them, which are not generally possible with synchrotron radiation experiments. In addition, the first step in the HHG is the tunneling ionization which is a highly nonlinear process and very selective with respect to the bonding energy of molecular orbital. Thus in general the HHG is dominated by the recombination to the highest-occupied molecular-orbital (HOMO) since it is usually the dominant channel for ionization due to its lowest bonding energy. However, the direct photoionization of a molecule is a linear process, where PICSs from inner molecular orbitals could be comparable to those from the HOMO. For instance, the HOMO of N_2 is a σ_g orbital. The next more tightly bound orbital, i.e., the HOMO-1 of N_2, is a π_u orbital which has an ionization energy 1.3 eV higher than the HOMO. The next following inner orbital, the HOMO-2 of N_2, is a σ_u orbital and is 1.9 eV more tightly bound than the HOMO-1. Both experimental results [7, 8] and calculations by Lucchese et al. [9] show that for the photon energy from the threshold at about 15 eV to about 40 eV, the PICS of HOMO-1 is actually comparable with that of HOMO while the HOMO-2 PICS is negligible, and around the photon energy of 29 eV, the PICS of HOMO has a shape resonance. These are experimental measurements and theoretical predictions made on randomly oriented N_2 molecules. It is interesting to know how these spectral features depend on the alignment of molecules. Actually there are rich structures in PICSs from fixed-in-space molecules predicted by the theory, however, they mostly remain unexplored experimentally. Can HHG spectra generated from aligned molecules provide the new information that are not yet directly available

C. Jin, *Theory of Nonlinear Propagation of High Harmonics Generated in a Gaseous Medium*, Springer Theses, DOI: 10.1007/978-3-319-01625-2_6,

from traditional photoionization measurements? The answer is yes. McFarland et al. [10] reported that the contribution of HOMO-1 to high harmonics dominates over that of the HOMO when molecules are aligned perpendicularly with respect to the polarization of HHG generating laser. Mairesse et al. [11] characterized the attosecond dynamics of multielectron rearrangement during the strong-field ionization by performing the harmonic spectroscopy.

Among the molecules, CO_2 is another most extensively studied system so far [12–15]. Initially the interest was focused on the observation of minimum in the HHG spectrum of CO_2 [2, 16, 17]. However, the positions of minima from different experimental measurements are often vastly different. PICSs of the HOMO for fixed-in-space CO_2 molecules indeed exhibit the minima at small alignment angles. According to the QRS theory, one can expect that the position of minimum in the harmonic spectrum does not significantly change with laser intensity assuming the HHG is generated from the HOMO only. Indeed, the strong-field ionization is very selective and depends exponentially on the ionization potential I_p. Since HOMO-1 and HOMO-2 orbitals in CO_2 are 4 and 4.4 eV more deeply bound than the HOMO, respectively, they are not expected to contribute significantly to the HHG. It is also well known that the tunneling ionization rates is very sensitive to the symmetry of molecular orbital [18]. For example, the HOMO of CO_2 is a π_g orbital, so ionization rates are small at small alignment angles. For the HOMO-2 of CO_2, it is a σ_g orbital, thus it has the large ionization rate when molecules are parallel aligned. Thus the HOMO-2 may become important to the harmonic generation even though it is bound 4.4 eV deeper than the HOMO at small alignment angles.[1] Theoretically, the alignment dependence of tunneling ionization rate is usually calculated by using the molecular Ammosov-Delone-Krainov (MO-ADK) theory [19, 20] or the strong-field approximation (SFA) [21]. Two models can give nearly identical alignment dependence (after normalization) for most molecules that have been studied so far.[2] However, this is not the case for CO_2. Experimentally, Pavičić et al. [23] reported the alignment dependent ionization of CO_2, and it was very narrowly peaked near the alignment angle of 46°. Their results differ significantly from the predictions of both MO-ADK and SFA [21]. In fact, this sharp alignment dependence reported in the experiment has not been confirmed by all theoretical attempts so far [24–28]. Furthermore, the reported experimental alignment dependence of ionization is also inconsistent with the observed HHG spectra from aligned CO_2 molecules [29, 30].

To extract the structure information from HHG spectra, earlier attempts were based on the SFA [31, 32] and the two-center interference model [33]. The quantitative rescattering (QRS) theory [34–36] was subsequently developed. It has been established that the HHG yield from an isolated molecule can be expressed as the product of a returning electron wave packet and a photorecombination cross section (PRCS). The QRS asserts that the PRCS is a free-field property of target and is

[1] The HOMO-1 of CO_2 is a π_u orbital, and it is not expected to contribute significantly to the HHG at small alignment angles, see Fig. 3 in [19] for the alignment dependent ionization.

[2] Usually, the relative ionization rate between different molecular orbitals cannot be expected the same from two models, one example of N_2 can be seen in [22].

independent of laser parameters, such as wavelength, intensity and pulse duration. The returning electron wave packet is determined by the laser solely. In the meanwhile, Smirnova et al. [2] reported HHG spectra from aligned CO_2 molecules and emphasized the role of hole dynamics playing in the HHG process with including the multiple orbitals. Multiple orbitals have also been easily incorporated into the QRS theory, this was first used by Le et al. [30] to explain the HHG data reported by McFarland et al. [10]. All of these theoretical studies were based on the single-molecule calculations. However, for high-harmonic generation, the phase-matching and macroscopic propagation effects in the medium cannot be avoided. For molecular targets, this has not been performed so far. Instead, it was often assumed that the HHG was measured under perfect phase-matching conditions and that the observed harmonics were directly proportional to the harmonics from a single molecule. While such assumptions may be inadequate if the accurate structure information of individual molecules is to be extracted from the observed HHG spectra.

In this chapter, I will first simulate measured HHG spectra of random and aligned N_2 molecules, and revisit the issue about the importance of HOMO-1 contribution raised in [10, 30, 37]. And then I will check how the famous shape resonance in the PICS of HOMO is presented in the HHG spectrum of N_2. Finally I will simulate measured HHG spectra of random and aligned CO_2 molecules, and investigate the effects of some factors which would change the position of minimum in the HHG spectrum of aligned CO_2 molecules. According to my knowledge, there have been no other theoretical attempts to simulate HHG spectra of molecular targets so far, which can be compared with measured ones directly. This chapter is organized as follows. In Sect. 6.2, I will show the calculated HHG spectra against measured ones for N_2 molecules being randomly distributed or aligned with the pump-probe angle as 0 degree, and the macroscopic wave packet extracted from calculated HHG spectra. In Sect. 6.3, I will discuss how the HOMO-1 contribution is displayed in the HHG spectrum of aligned N_2 molecules with the pump-probe angle of 90 degrees as laser intensity varies. In Sect. 6.4, I will show the PICS of N_2 from the well-established theory, and the shape resonance in the HHG spectrum at small alignment angles. In Sect. 6.5, I will simulate existing HHG spectra of CO_2 molecules, and discuss the origin of minimum for aligned cases. In Sect. 6.6, I will investigate carefully how the minimum in the HHG spectrum of aligned CO_2 shifts its position. A short summary in Sect. 6.7 will conclude this chapter. Note that in this chapter I am only concerned of the parallel component of HHG along the polarization direction of generating laser.

6.2 HOMO Contribution in HHG of Random and Aligned N_2

6.2.1 *Macroscopic HHG Spectra: Theory Versus Experiment*

HHG spectra by 1200-nm lasers have been reported by Wörner et al. [38] for randomly distributed and aligned N_2 molecules recently. In this section, I show the simulated results for N_2 molecules, at two laser intensities, 0.9 and 1.1×10^{14} W/cm^2, reported in [38]. Experimentally, laser duration (full width at half maximum, FWHM) is $\sim$44 fs, beam waist at the focus is $\sim$40 μm, and gas jet is located 3 mm after laser focus with the length of $\sim$1 mm. A vertical slit with a width of 100 μm is placed 24 cm after the gas jet to mostly select "short" trajectories. To achieve the good agreement in the cutoff, laser intensities (in the center of the gas jet) used in the theoretical simulations are 0.78 and 0.9×10^{14} W/cm^2 instead. Since the experiment was performed at low laser intensity and low gas pressure, the fundamental laser field is not modified through the medium as discussed in Sect. 2.3.3, and high harmonics are propagated without absorption and dispersion effects from the medium. In the theoretical simulation, induced dipoles of fixed-in-space molecules are obtained by using the QRS theory (as discussed in Sect. 2.2.3) for different laser intensities, they are averaged coherently according to the alignment distribution, see Eq. (2.5), and then fed into the Maxwell's wave equations.

In Fig. 6.1, the good overall agreement between measured and simulated spectra are shown, for both randomly distributed and aligned N_2 molecules. In experimental HHG spectra, as seen from Fig. 1 b and c in [38], there are shallow minima at 38 $\pm$ 2 eV (at low intensity) and at 41 $\pm$ 2 eV (at high intensity) for both aligned and unaligned molecules. The theory also predicts a minimum in the HHG spectrum. For random molecules, the minimum is at $\sim$39 eV for low intensity and $\sim$40 eV for high intensity. For aligned molecules, the minimum is at $\sim$42 eV for low intensity and $\sim$44 eV for high intensity. The alignment degree in the experiment was estimated to be $\langle \cos^2 \theta \rangle = 0.6 - 0.65$. In the simulation, $\cos^4 \theta$ is used to describe the alignment

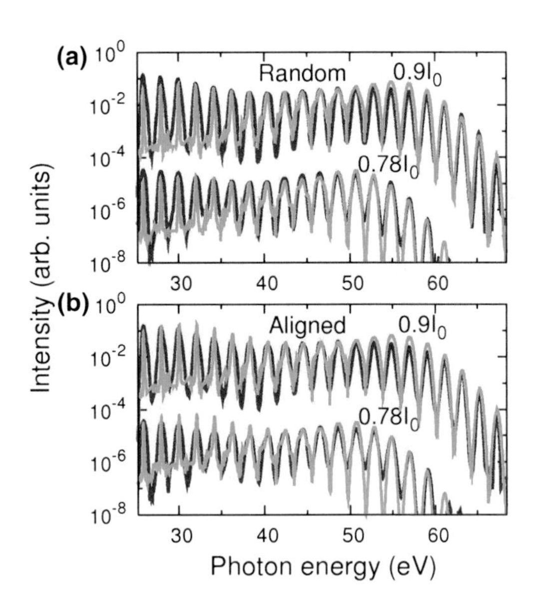

Fig. 6.1 Comparison of HHG spectra from theory (*green curves*) and experiment (*red curves*) in [38], **a** for randomly distributed N_2 molecules and **b** for N_2 molecules aligned along the polarization direction of generating laser, i.e., the pump-probe angle $\alpha = 0°$. Laser intensities are indicated with $I_0 = 10^{14}$ W/cm^2. See text for additional laser parameters. Adapted from [39].

distribution for simplicity.[3] Note that only the HOMO of N_2 molecules is included in the calculation. As far as I can judge, this is the first time that simulated HHG spectra of molecules have been compared directly to the measured spectra, where phase-matching and propagation effects in the medium are included in the simulations.

6.2.2 Separation of PR Transition Dipole from HHG Spectra

As shown in Sect. 3.5, the macroscopic HHG spectrum for atomic targets is $S_h(\omega) \propto \omega^4|W'(\omega)|^2|d(\omega)|^2$, where $W'(\omega)$ is called a "macroscopic wave packet" (MWP), and $d(\omega)$ is the PR transition dipole moment. For the macroscopic HHG spectrum of partially aligned molecules, this equation is still valid

$$S_h(\omega) \propto \omega^4|W'(\omega)|^2|d^{avg}(\omega, \alpha)|^2, \tag{6.1}$$

with $d(\omega)$ being replaced by coherently averaged PR transition dipoles[4]:

$$d^{avg}(\omega, \alpha) = \int_0^{2\pi}\int_0^{\pi} N(\theta')^{1/2} d(\omega, \theta')\rho(\theta', \phi', \alpha)\sin\theta' d\theta' d\phi', \tag{6.2}$$

where $N(\theta')$ is the alignment-dependent ionization probability, $\rho(\theta', \phi', \alpha)$ the alignment distribution defined in Eq. (2.59), and $d(\omega, \theta')$ the parallel component of alignment-dependent PR transition dipole. Here θ' and ϕ' are alignment angles with respect to the polarization direction of probe laser, and α is the pump-probe angle. From Eq. (6.1), the target structure is reflected in the PR transition dipole, and the phase-matching and propagation effects of high harmonics, in the meanwhile, are incorporated into the MWP. The MWP also can be considered as the cumulative effect of the returning electron wave packet (or the microscopic wave packet) after the propagation in the medium and in free space. The validity of Eq. (6.1) forms the basis of extracting the molecular structure information from experimentally measured HHG spectra.

For HHG spectra shown in Fig. 6.1, one can separate averaged PR transition dipoles and MWPs according to Eq. (6.1), see Fig. 6.2. The averaged PR transition dipole indeed shows a rapid drop near 40 eV, which is due to the presence of a shape resonance of N_2 around 30 eV, and it is more pronounced for aligned molecules than for random ones.[5] I have checked that the MWP for randomly distributed molecules is the same as that for aligned molecules under the same laser intensity. Thus it explains why Le et al. [5] could use the HHG from single-molecule response only to interpret how the yield of each harmonic changes with the pump-probe time delay. However,

[3] In principle the alignment distribution can be calculated by solving the TDSE, will be implemented in later sections.

[4] In [22, 40], the integral over ϕ' in Eq. (6.2) is incorporated into $\rho(\theta', \alpha)$.

[5] The origin can be seen from PICSs in Fig. 6.5.

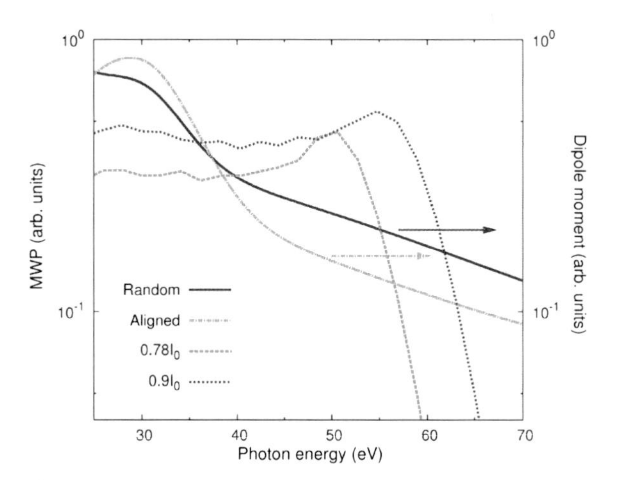

Fig. 6.2 Macroscopic wave packets (MWPs) for two different laser intensities, and averaged PR transition dipoles are for random and aligned N_2 molecules. MWPs are extracted from HHG spectra in Fig. 6.1 according to Eq. (6.1). Adapted from [39].

the MWP greatly changes with laser intensity, especially for the longer wavelength laser used here. As shown in Fig. 6.2, two MWPs have somewhat different slopes. Thus the multiplication of the averaged PR transition dipole and the MWP results in a weak minimum in the observed HHG spectrum. Comment that structures in the PR transition dipole[6] could be severely averaged out when molecules are not well aligned, and the minimum in the HHG spectrum would be more clearly seen if molecules were better aligned. The position of minimum could be shifted a little bit. Furthermore, MWPs in Figs. 3.6 and 6.2 are rather different due to the large difference in laser intensities used. In the future, it is desirable that the predictions such as those in Figs. 3.6 and 6.2 be checked under the same experimental conditions carefully.

6.3 Intensity Dependence of Multiple Orbital Contributions in HHG of Aligned N_2

6.3.1 Macroscopic HHG Spectra: Theory Versus Experiment

In Sect. 6.2, I have simulated HHG spectra of aligned N_2 molecules with the pump-probe angle $\alpha = 0°$. In this section, I would like to simulate new measurements for aligned N_2 molecules at a time delay corresponding to the maximal alignment while the HHG-generating laser is perpendicular to the aligning one ($\alpha = 90°$). The degree of alignment is estimated to be $\langle \cos^2\theta \rangle = 0.60 \pm 0.05$.

In Fig. 6.3, it shows the simulated HHG spectra of aligned N_2 molecules generated by a 1200-nm laser which is perpendicular to the aligning laser. Experimental mea-

[6] See Fig. 6.5 for reference.

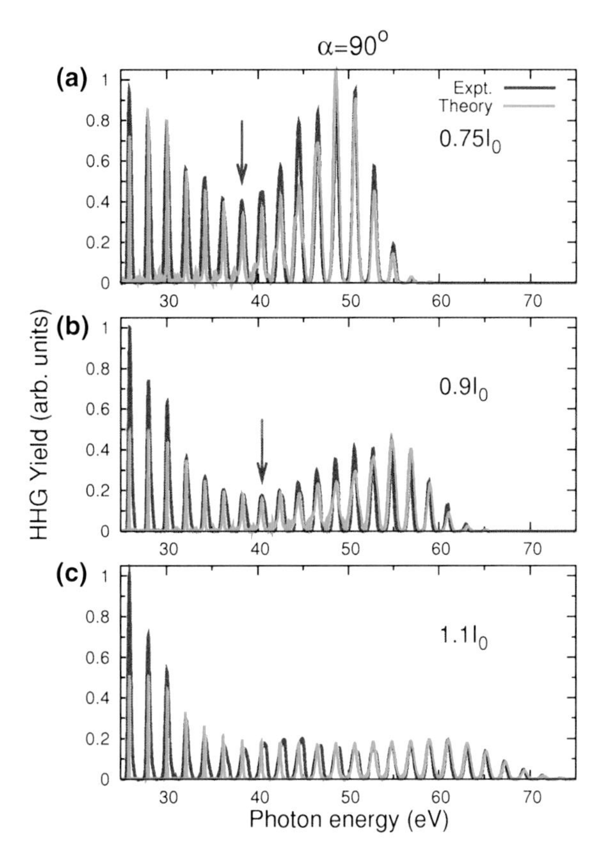

Fig. 6.3 Comparison of experimental (*red lines*) and theoretical (*green lines*) HHG spectra of aligned N_2 molecules in a 1200-nm generating laser where the pump-probe angle $\alpha = 90°$. Laser intensities in the simulations are indicated with $I_0 = 10^{14}$ W/cm^2. Alignment degree is $\langle \cos^2 \theta \rangle = 0.60$ by the pump laser. See text for additional laser parameters. In **a** and **b**, only σ orbital is included in the simulations, and in **c** both σ and π orbitals are included in the simulation. *Arrows* indicate the positions of minima. Adapted from [22]. © (2012) by the American Physical Society

surements are also shown. Experimental conditions are the same as those in Sect. 6.2. Laser intensities (in the center of the gas jet) used in the theory are adjusted to coincide with the position of experimental HHG cutoff. In Fig. 6.3a–c, laser intensities in theory (experiment) are 0.75 (0.65), 0.9 (1.1) and 1.1 (1.3), in units of 10^{14} W/cm^2, respectively. In the simulation, the degree of alignment is $\langle \cos^2 \theta \rangle = 0.60$,[7] and other parameters are chosen close to the experimental ones. HHG spectra from the theory and the experiment are normalized at the cutoff.

Main features in the spectra are two deep minima (as indicated by the arrows) at 38.2 and 40.4 eV, at two low laser intensities in Fig. 6.3a and b. In Fig. 6.3c, the minimum disappears at the highest laser intensity. To simulate the spectra at two low intensities, only the HOMO (σ orbital) is included. In Fig. 6.3a and b, both correct shapes of the spectra and the precise positions of minima in spectra are reproduced by the simulation. For the highest intensity in Fig. 6.3c, the theory could not reproduce the correct spectral shape when only the σ orbital is included. It also predicted a minimum in the HHG spectrum which was not seen in the experiment, see Fig. 6.4a.

[7] See Appendix B for the quantum calculation of molecular alignment.

A very good agreement between the theory and the experiment (correct shape and no minimum in the spectrum) in Fig. 6.3c is then achieved if both σ and π (HOMO-1) orbitals of N_2 are included.

6.3.2 Single Orbital (HOMO) Contribution at Low Laser Intensity

I first take a careful analysis of the spectral features in Fig. 6.3a and b. The deep minimum in the HHG spectrum of N_2 molecules is related to the σ orbital, which has been observed in many experiments, either in unaligned or aligned N_2 [17, 38, 41, 42]. When laser intensity is changed, this minimum shifts only slightly. This behavior is quite similar to the well-known Cooper minimum in the HHG spectrum of Ar [39, 43, 44]. The same behavior, i.e., the position of minimum slightly shifting with the laser intensity, has been observed by Wörner et al. [38] (see their Fig. 1) and Farrell et al. [42] (see their Fig. 7) when $\alpha = 0°$. This can be understood as proposed in Sect. 6.2 in terms of the concept of MWP. When only one molecular orbital dominates the harmonic generation, the harmonic yield can be expressed in Eq. (6.1). In Fig. 6.4b, it shows a fast drop-off in the averaged PR transition dipole (the square of the magnitude multiplied by ω^2 is shown in the figure) of the σ orbital around 38 eV, which causes the pronounced minimum in the HHG spectrum in Fig. 6.3a, b and Fig. 6.4a. Note that the slope of averaged PR transition dipole does not change, but the MWP ($\omega^2|W'(\omega)|^2$) increases monotonically with the photon energy for different laser intensities. When laser intensity at the focus is increased, the spatial distribution of infrared (IR) laser intensity is different, so the wave packet is modified differently as it propagates through the medium, see Fig. 6.2. As discussed in Sect. 6.2.2, this makes the minimum in the HHG spectrum shift slightly with the laser intensity. With only one molecular orbital, the position of minimum in the experiment can be located very close to the one in the theoretical simulation.

6.3.3 Multiple Orbital (HOMO and HOMO-1) Contributions at Higher Laser Intensity

I next examine the spectral features in Fig. 6.3c. Figure 6.4a shows the envelopes of HHG spectra in Fig. 6.3c from two individual molecular orbitals, and from the total one, obtained after the macroscopic propagation in gas medium and in free space. Meanwhile, the averaged PR transition dipoles (the square of the magnitude multiplied by ω^2, the degeneracy is not included) of two molecular orbitals are shown in Fig. 6.4b.

In Fig. 6.4a, the σ orbital alone shows a deep minimum. The interference between σ and π orbitals washes out the minimum in the spectrum because these two orbitals show comparable contributions over the whole spectral region. Based on the QRS

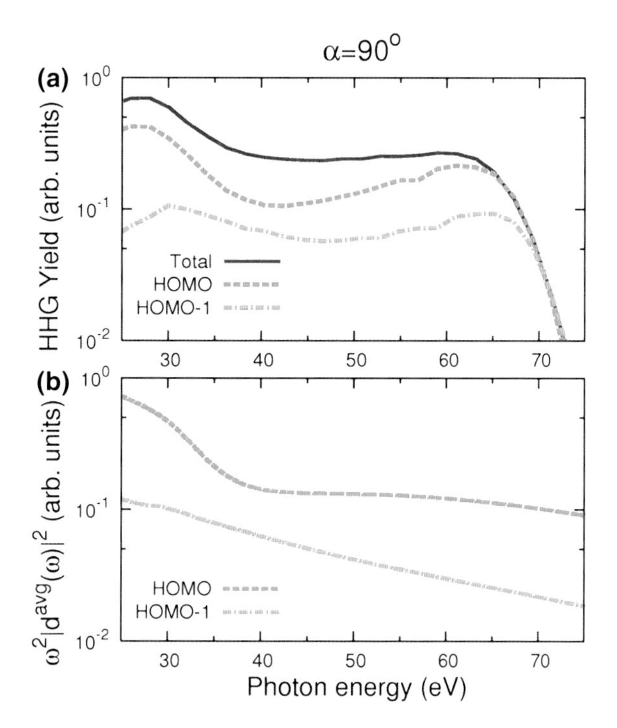

Fig. 6.4 **a** Calculated macroscopic HHG spectra (envelope only) corresponding to Fig. 6.3c. Total (HOMO and HOMO-1 together) spectra and individual HOMO and HOMO-1 spectra are plotted. **b** Averaged photorecombination transition dipoles (parallel component, the square of the magnitude multiplied by ω^2) of HOMO and HOMO-1 corresponding to **a**. Laser intensity is 1.1×10^{14} W/cm^2. Alignment degree is $\langle \cos^2 \theta \rangle = 0.60$ by the pump laser. Pump-probe angle is $\alpha = 90°$. Adapted from [22]. © (2012) by the American Physical Society

theory, macroscopic HHG spectra from individual molecular orbitals can be considered as a product of a MWP and an averaged PR transition dipole as shown in Eq. (6.1). Since the ionization potential of π orbital (16.9 eV) differs from σ orbital (15.6 eV) only by 1.3 eV, the MWPs of two orbitals are almost same under the same IR laser, so the relative contribution between σ and π orbitals to the total HHG spectrum is mostly determined by the averaged PR transition dipoles. Furthermore, the relative contribution between two orbitals can be adjusted by alignment-dependent ionization probabilities $N(\theta')$ and the alignment distribution $\rho(\theta', \phi', \alpha)$ as involved in Eq. (6.2). In Fig. 6.3a and b, the magnitude of $N(\theta')$ for σ orbital is much larger than the one for π orbital, thus making the corresponding averaged PR transition dipole also larger. At higher intensity of Fig. 6.3c, the relative magnitude of $N(\theta')$ for π orbital becomes bigger, the averaged transition dipoles between two orbitals become comparable. Thus both orbitals contribute to the HHG and the interfere results in a drastic change of the spectra. One can see that the total HHG can evolve from single-orbital to the multiple-orbital phenomena by simply increasing the laser intensity. Note that the $N(\theta')$ in this section is calculated by using the MO-ADK theory [19, 20]. There are also other methods or models in the literature [27, 30, 45] for calculating molecular ionization rates. Theoretical predictions may be changed by using different ionization rates or different alignment distributions. I find that ionization rates for a few molecular orbitals obtained from the MO-ADK theory are very close to the model calculations by Petretti et al. [27] where they solved the time-dependent Schrödinger equation (TDSE) at laser intensity of 1.5×10^{14} W/cm^2.

The issue of π orbital contribution to the HHG of N_2 molecules when the pump-probe angle $\alpha = 90°$ has been addressed previously [10, 37] using traditional 800-nm pulses at high laser intensities of around 2.0×10^{14} W/cm^2. The HOMO-1 was found to be much more pronounced in the cutoff region (or beyond the classical cutoff). Note that with the ionization probability obtained from the SFA [30] the QRS theory has been applied to interpret results in [10] without including phase-matching and propagation effects.[8] In this section, a longer-wavelength (1200 nm) laser generates a much broader photon-energy range even with a low laser intensity, and π orbital contributes not only in the cutoff, but also in the plateau.

In the future, experimental measurements with a full range of pump-probe angles may provide a way to determine the relative ionization probabilities of two orbitals. If one wishes to probe the PR transition dipole of HOMO at the pump-probe angle $\alpha = 90°$, the HHG spectrum taken at a low laser intensity with a long-wavelength laser is preferable to avoid the multiple orbital contributions because the pump-probe angle $\alpha = 90°$ is much closer to the alignment angle of 90° in considering the alignment distribution in different frames. This would make the retrieval of target structure easier. To isolate the π orbital at the pump-probe angle $\alpha = 90°$, a higher laser intensity and better alignment are required to greatly enhance its contribution to the HHG [10, 37].

6.4 Shape Resonance in Photoionization and HHG of N_2

6.4.1 PICSs and Phases from HOMO and HOMO-1 Orbitals

Parallel and perpendicular amplitudes of the transition dipole in the body-frame of molecule are the most basic information on the photoionization of a fixed-in-space molecule. They are involved in the differential PICS in the body-fixed frame:

$$\frac{d^2\sigma}{d\Omega_{\hat{k}} d\Omega_{\hat{n}}} = \frac{4\pi^2\omega k}{c} \mid \langle \Psi_i | \vec{r} \cdot \hat{n} | \Psi^{(-)}_{f,\vec{k}} \rangle \mid^2, \tag{6.3}$$

where ω is the photon energy, $\hat{n}$ the polarization direction of light, and $\vec{k}$ the momentum of the photoelectron. I will only focus on the case of $\hat{n} \parallel \vec{k}$ since this is related to the parallel polarized HHG component measured from aligned molecules.

Figure 6.5 shows the differential PICSs (in units of Mb) and the corresponding phases of σ and π orbitals of N_2 molecules for the photon energy from 20 to 80 eV by

[8] At the alignment angle of 90°, the ratio of ionization rates between the HOMO and the HOMO-1 is about 2 from the SFA in comparison with about 5 from the MO-ADK theory [19, 20].

using the well-established photoionization theory.[9] Above observed HHG minima as shown in Fig. 6.3 for $\alpha = 90°$ and in Fig. 6.1 for $\alpha = 0°$ and randomly distributed N_2 molecules can all be attributed to the rapid change of HOMO cross section near 40 eV at alignment angles either close to 0° or 90°. However, the precise location of minimum depends on the alignment degree. For HOMO-1, the cross section generally peaks at large alignment angles. Thus the interference of HHG spectra from σ and π orbitals can only be observed close to $\alpha = 90°$ (this is equivalent to the large alignment angle). PICSs in Fig. 6.5 for two orbitals are shown on the same scale. The PICS from the HOMO-1 is comparable with that from the HOMO over photon energies covered except for the HOMO shape resonance near 30 eV that will be discussed next. In the HHG, the HOMO contribution is always dominant for randomly distributed N_2 and for aligned N_2 at small alignment angles and lower laser intensities, while the HOMO-1 contribution can become more important only at large alignment angles and higher laser intensities, as discussed earlier.

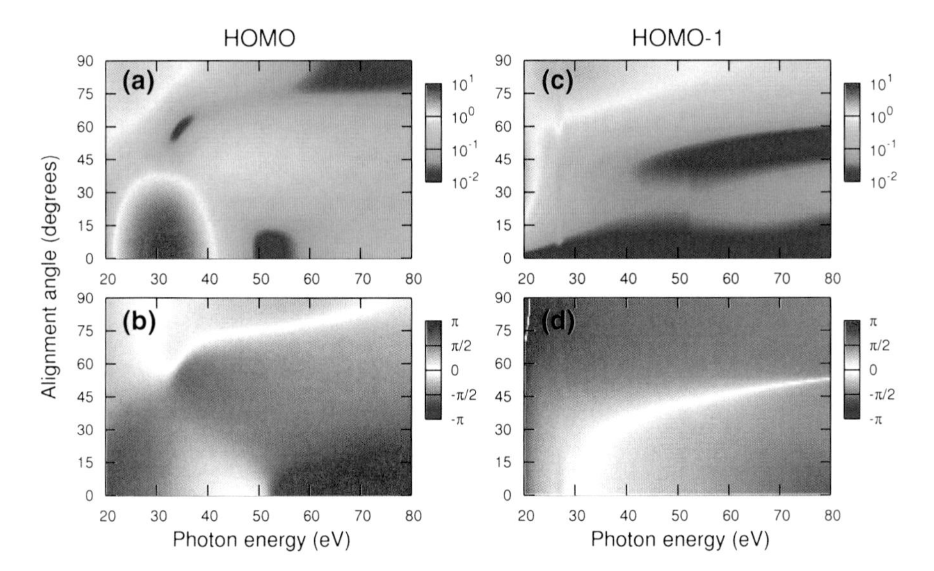

Fig. 6.5 Calculated differential photoionization cross sections (**a** and **c**) and phases (**b** and **d**) of HOMO and HOMO-1 for N_2 molecules in terms of alignment angles, respectively. Shape resonance in the HOMO shows up only at small alignment angles around 30 eV. Only the parallel component to the polarization direction of laser is shown. Adapted from [22]. © (2012) by the American Physical Society

[9] See Appendix C.

6.4.2 Shape Resonance in HHG of Aligned N_2

Most of resonances which are very common in the photoionization are due to so-called Feshbach resonances. In general they can only be observed using the high-resolution spectroscopy since they are quite narrow. However, broad Feshbach resonances and shape resonances have widths from the fractions to a few eV's. In the HHG, resonances have been recently explored in the theory [46–49] and experiment [50] for atomic targets. The resonance feature observed in Xe [50] is due to the inter-shell coupling with the well-known shape resonance that occurs in $4d$ shell. But shape resonances are quite rare for common atomic targets. On the other hand, shape resonances are very common in molecules. For example, there is a pronounced shape resonance in the HOMO channel of N_2 near the photon energy of 30 eV. This resonance for small alignment angles only is caused by $3\sigma_g \rightarrow k\sigma_u$ channel. In Fig. 6.5b, one can see that there is a decrease of the phase shift by π from 20 to 40 eV for this resonance. Note that this is the only known shape resonance in the covered energy region. To observe this shape resonance in the HHG, clearly it is better to select the ionization from σ orbital only by using low laser intensity and for molecules that are aligned nearly parallel to the polarization axis of probing laser.

In Fig. 6.6, the calculated HHG spectra (envelope only, normalized, and after the macroscopic propagation) of randomly distributed N_2 molecules at two laser

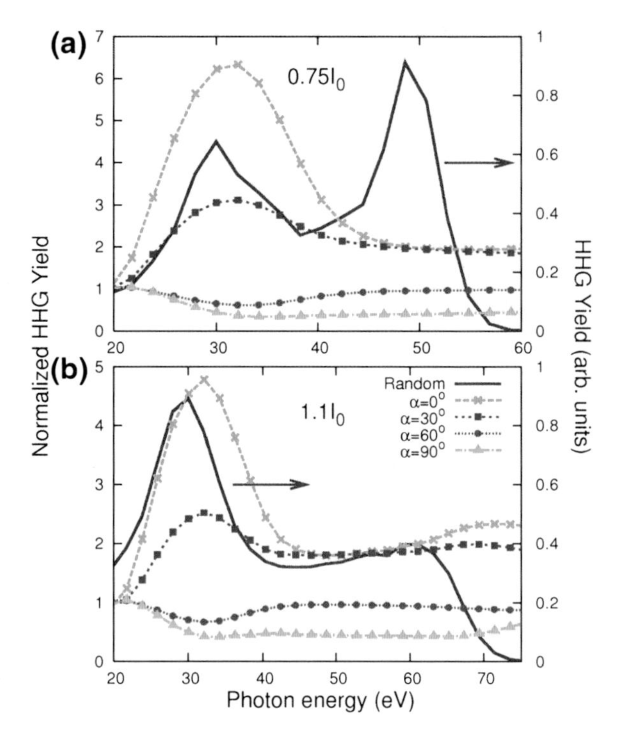

Fig. 6.6 Calculated macroscopic HHG spectra (envelope only) of unaligned N_2 and normalized HHG yields of aligned N_2 with respect to unaligned ones as a function of the photon energy at selected pump-probe angle α. Laser intensities are indicated with $I_0 = 10^{14}$ W/cm^2. Alignment degree is $\langle \cos^2 \theta \rangle = 0.60$. Adapted from [22]. © (2012) by the American Physical Society

intensities are shown. For randomly distributed N_2, it is generally known that HOMO is dominant in the harmonic generation. High harmonics peak around 30 eV in the spectra are due to the shape resonance in the PICS of σ orbital at small alignment angles. In Fig. 6.6, the normalized harmonic yield with respect to the randomly distributed one at selected alignment angles is shown. The degree of alignment is taken as $\langle\cos^2\theta\rangle = 0.60$. I obtain the intensity of each odd harmonic q by integrating the harmonic yield over harmonics $q-1$ to $q+1$. The shape resonance is very pronounced at $\alpha = 0°$ for low laser intensity. As the pump-probe angle is increased, the shape resonance decreases, showing that it is present only at small aligning angles. For higher laser intensities, the same behavior is seen even though π orbital also contributes at large pump-probe angles. Comment that the absorption which is not included in the present simulation would suppress the shape resonance if gas pressure is high.

It is far from conclusive if one compares the above prediction with existing experimental data. Lee et al. [37] reported the HHG ratios of aligned versus unaligned N_2 at selected alignment angles using 800-nm lasers. They used a high laser intensity of 2.5×10^{14} W/cm^2, see their Fig. 1(c), but they did not present data near the resonance region. My calculation in this chapter does not reproduce their measured ratios. Using a 1300-nm laser, Torres et al. [17] have shown the high-harmonic data for aligned N_2 with intensity of 1.3×10^{14} W/cm^2, see their Fig. 4. The general trend of their data is very close to the prediction in this chapter, but they used a higher intensity and a higher degree of alignment ($\langle\cos^2\theta\rangle = 0.66$), and thus the HOMO-1 channel may contribute to the signal at larger alignment angles. The experimental HHG measurement by Kato et al. [51] did not extend the photon energy below 30 eV either. There are other measurements in the literature [42, 52, 53] using Ar or aligned N_2 at $\alpha = 0°$ as a reference. It is difficult to directly compare with these data. Thus, it remains interesting to check if the shape resonance in the PICS of N_2 can be seen in the HHG spectrum as predicted here. Comment that the absorption was not included in the propagation simulation, which may modify the prediction if gas pressure is too high. Experiments dedicated to address this issue would be of interest.

6.5 Contributions of Multiple Molecular Orbitals in HHG of Aligned CO_2

6.5.1 Macroscopic HHG Spectra: Theory Versus Experiment

HHG spectra for isotropically distributed and partially aligned CO_2 molecules by 800-nm and 1200-nm lasers have been measured in [38]. In this section, I will present the simulated HHG spectra for random and aligned CO_2 molecules in the following. The SFA is used to calculate the ionization probability in the QRS theory in this section and next section, unless otherwise stated.

1. HHG Spectra of Randomly Distributed CO_2

In Fig. 6.7a–c, HHG spectra for isotropically distributed CO_2 molecules by 800-nm and 1200-nm lasers are shown. Laser intensity used in the simulation is adjusted from the value given in the experiment to obtain the good agreement between theory and experiment, especially in the cutoff. Laser intensities in the theory (experiment) are 1.9 (2.1), 0.8 (1.0) and 1.0 (1.2), in units of 10^{14} W/cm^2, respectively. Other parameters used in the theory are chosen close to those given in the experiment [38]. Laser parameters are that pulse duration is ~32 fs (800 nm) or ~44 fs (1200 nm), and beam waist at the focus is ~40 μm. A 0.6-mm-wide gas jet is located 3 mm (800 nm) or 3.5 mm (1200 nm) after laser focus, and a slit with a width of 100 μm is placed 24 cm after gas jet.

In Fig. 6.7a–c, they clearly show the good overall agreement between experiment and theory for randomly distributed CO_2 molecules. I have checked that with the negligible contributions from inner molecular orbitals, HOMO is dominant for randomly distributed CO_2. CO_2 spectra here do not exhibit any minima, as opposed to

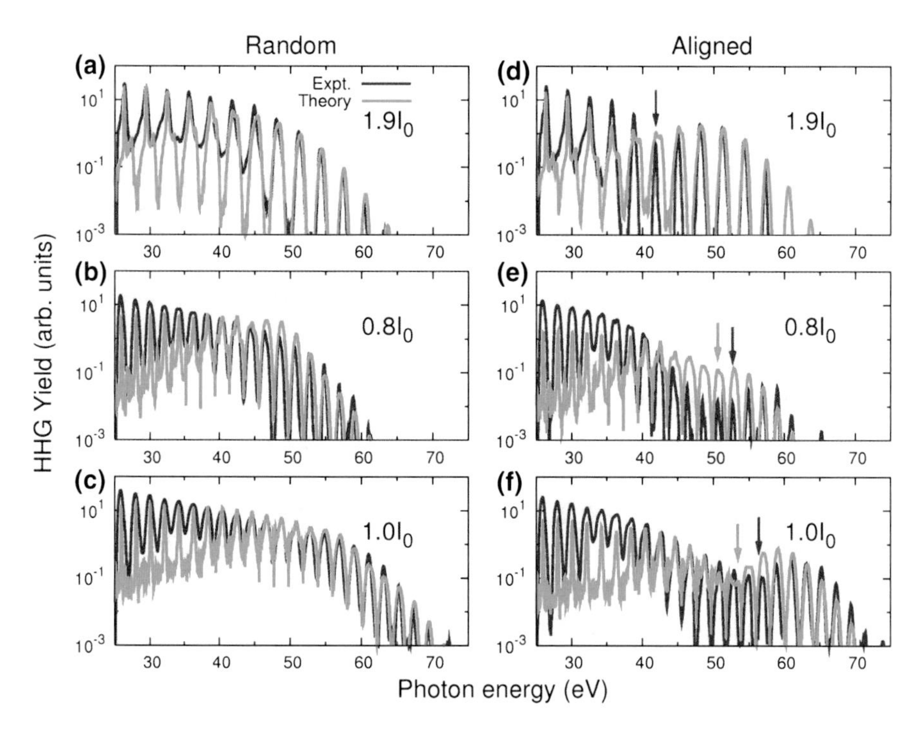

Fig. 6.7 Comparison of theoretical (*green lines*) and experimental (*red lines*) HHG spectra of random and aligned CO_2 molecules, in an 800-nm laser shown in **a** and **d**, and in a 1200-nm laser shown in **b**, **c**, **e** and **f**. Laser intensities are indicated with $I_0 = 10^{14}$ W/cm^2. Experimental data are from [38]. Pump-probe angle is $\alpha = 0°$. *Arrows* indicate the positions of minima. See text for additional laser parameters and experimental arrangements. Adapted from [40]. © (2011) by the American Physical Society

the spectra of random N_2 molecules as shown in Fig. 6.1 when HHG spectra from these two molecules are generated under the similar experimental conditions [38]. For randomly distributed CO_2 molecules, Vozzi et al. [16] have reported that there was no minimum in the HHG spectrum using an 800-nm laser. However, HHG data from Torres et al. [17, 54] appeared to show a weak minimum at the photon energy near 45 eV for 1300-nm lasers. Without more careful studies including different intensities and wavelengths, however, this is not conclusive.

2. HHG Spectra of Aligned CO_2

Experimentally, molecules could be impulsively aligned by using a relatively weak and short laser pulse, and HHG spectra were taken at the half-revival ($\sim$21.2 ps in CO_2) when molecules were maximally aligned [38]. By solving the TDSE of rotational wave packet,[10] one could obtain angular distributions of aligned molecules. In the simulation, the degree of alignment is assumed as $\langle \cos^2 \theta \rangle = 0.60$ in Fig. 6.7d, and $\langle \cos^2 \theta \rangle = 0.50$ in Fig. 6.7e, f. The polarization of pump laser is parallel to that of probe laser.

In Fig. 6.7d and f, HHG spectra of partially aligned CO_2 molecules are shown. These spectra are generated under the same probe lasers and experimental arrangements as those in Fig. 6.7a–c. Both HOMO and HOMO-2 of CO_2 contribute to the HHG. The simulations and the experimental data are in the general agreement. In Fig. 6.7e, the discrepancy between experiment and theory is a little bigger. The drop in the spectrum near 40 eV is smaller in the theory than that in the experiment. But the experimental data do not drop so rapidly and are in the agreement with the theoretical simulation in Fig. 6.7f.

The minima in HHG spectra of CO_2 molecules and their dependence on laser intensity have been widely studied in the literature [54, 2]. For an 800-nm laser, in Fig. 6.7d, the experiment gives a strong minimum at 42 ± 2 eV, the simulation predicts a minimum around 42 eV. In Fig. 6.7e, the experiment shows a minimum at 51 ± 2 eV, the theory predicts a minimum around 50 eV for the 1200-nm laser. The experimental minimum is then shifted to 57 ± 2 eV in Fig.6.7f, and the theoretical one is accordingly moved to around 53.5 eV. Thus as laser intensity is increased the simulation also shows the shift of minimum from low to high harmonic orders.

6.5.2 Origin of Minimum in the HHG Spectrum of Aligned CO_2

Here I analyze the origin of minimum in the HHG spectrum observed in Fig. 6.7d–f, and consider dominant contributions from HOMO and HOMO-2 only. The averaged PR transition dipole is defined in Eq. (6.2) for each molecular orbital, which is obtained by averaging over the alignment distribution of partially aligned molecules, weighted by the square root of the tunneling ionization probability. Thus the relative

[10] See Appendix B.

contribution of each molecular orbital to the HHG can be measured. Note that the relative ionization rates between HOMO and HOMO-2 change with laser intensity as discussed in Sect. 6.3.3 for N_2 molecules.

Envelopes of HHG spectra from individual molecular orbitals together with the total ones are shown in Fig. 6.8a–c, each obtained after the propagation in the medium. In the meanwhile, Fig. 6.8d and f show averaged PR transition dipoles of HOMO and HOMO-2 under different generating lasers and alignment distributions.

In Fig. 6.8a and b, HHG spectra of HOMO or HOMO-2 don't have any minima, but the minimum shows up in the total spectrum. The interference between the HOMO and the HOMO-2 leads to this minimum. This is called as type I minimum. It is clearly seen that as laser intensity is changed the minimum position will be varied due to the relative ionization rates between HOMO and HOMO-2 changing with the intensity, also see Fig. 6.11c and d. The similar analysis was presented in [2]. In Fig. 6.8c, the HOMO spectrum has a minimum at 52.6 eV. Due to the interference with the HOMO-2, this minimum is shifted to 53.6 eV in the total spectrum. This is categorized as type II minimum. The similar analysis can be found in [54, 55]. The minimum in the HOMO spectrum is caused by the minimum in the averaged PR transition dipole of HOMO as shown in Fig. 6.8f. Due to the modification of

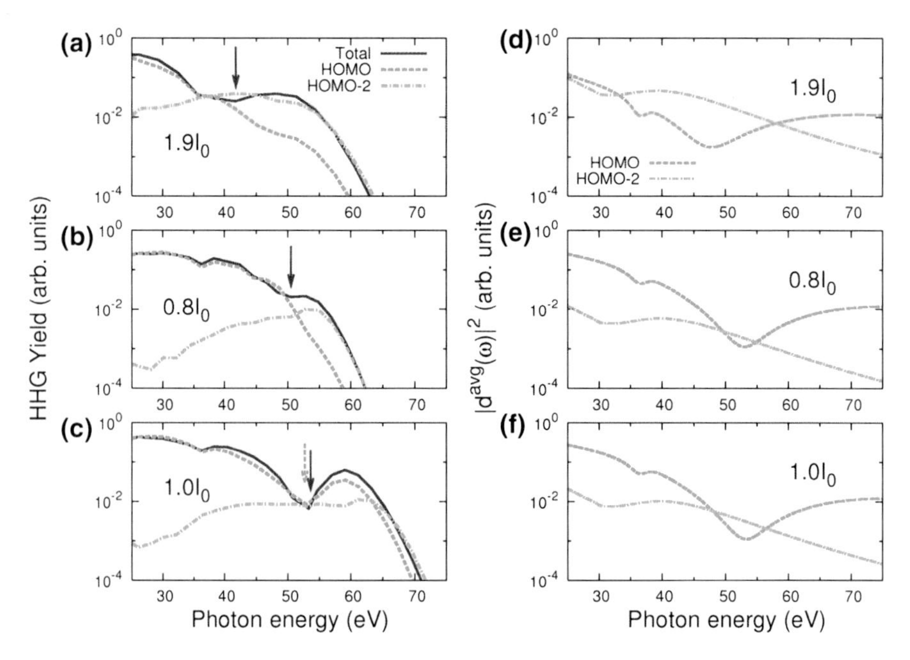

Fig. 6.8 **a**–**c** Macroscopic HHG spectra (envelope only) corresponding to Fig. 6.7**d**–**f**, respectively. Total (HOMO and HOMO-2 together) spectra and the spectra of individual HOMO and HOMO-2 are shown. **d**–**f** Averaged photorecombination transition dipoles (parallel component, the square of magnitude) of HOMO and HOMO-2 corresponding to **a**, **b** and **c**. Laser intensities are indicated with $I_0 = 10^{14}$ W/cm^2. Pump-probe angle is $\alpha = 0°$. *Arrows* indicate the positions of minima. Adapted from [40]. © (2011) by the American Physical Society

"macroscopic wave packet" (MWP), its position differs from each other sometimes. With the decreased degree of alignment, the minimum in the averaged PR transition dipole from the HOMO in Fig. 6.8d is slightly shifted to the higher energies, see Fig. 6.8 e and f. Furthermore, as shown in Fig. 6.2, the minimum in the HHG spectrum of HOMO could also result from the multiplication of averaged PR transition dipole and MWP, even when neither has the minimum. One can expect that when a minimum occurs in the dominant orbital and this orbital remains the dominant one as laser intensity is changed, the position of minimum will not change much. There are few bumps in the HHG spectra of HOMO around 36 eV as well as in total spectra, which can be seen in Fig. 6.8 d and f due to the bumps in averaged PR transition dipoles of HOMO. Their positions do not change much since the HOMO remains dominant.

As shown in Fig. 6.2, the macroscopic HHG spectrum is the product of an averaged PR transition dipole and a MWP for each individual molecular orbital. The ionization rate (or ionization probability) for each molecular orbital has been incorporated in the averaged PR transition dipole, so the MWP is mostly identical except for the phase due to ionization potential. The averaged PR transition dipole is sensitively dependent on the ionization rate. As laser intensity is increased, the relative magnitude of two averaged PR transition dipoles (from different molecular orbitals) changes rapidly. Thus when they are comparable, see Fig. 6.8d, the position of minimum changes rapidly with laser intensity accordingly. The alignment distribution is another important factor to compose the averaged PR transition dipole, see Fig. 6.8 d and f. At low laser intensity, due to the small contribution from HOMO-2, the interference often occurs in a narrow region only where two contributions from HOMO and HOMO-2 are comparable, see Fig. 6.8 b and c. In comparison, HOMO and HOMO-2 tend to interfere over a broad photon-energy region in Smirnova et al. [2]. They also used different ionization rates and transition dipoles.

6.6 Major Factors that Influence the Positions of Minima in HHG Spectra of Aligned CO_2

6.6.1 Progression of Harmonic Minimum Versus Laser Intensity

In Fig. 6.9a and b, the envelopes of calculated HHG spectra for four different laser intensities with an 800-nm laser and a 1200-nm laser are shown. For 800-nm spectra, there is no minimum for the lowest laser intensity. For higher intensities, each spectrum has a type I minimum, with its position shifting to the higher photon energy as laser intensity is increased. The degree of alignment of molecules in the calculation is chosen as $\langle\cos^2\theta\rangle = 0.60$. I find that the minimum shift cannot be attributed to either the averaged PR transition dipole or the MWP alone. For 1200-nm data, I find that the minimum is type II, where the averaged PR transition dipole of HOMO has a minimum, with the alignment degree $\langle\cos^2\theta\rangle = 0.60$, which is different from that in Fig. 6.7e and f. As laser intensity increases, the minimum in the HHG spectrum of

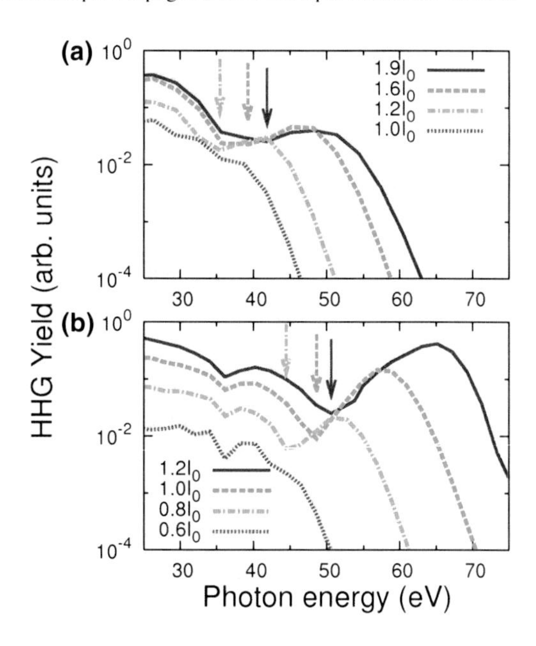

Fig. 6.9 Laser intensity dependence of macroscopic HHG spectra (envelope only) **a** in an 800-nm laser and **b** in a 1200-nm laser. Intensities are shown in units of $I_0 = 10^{14}$ W/cm^2. Degree of alignment: $\langle \cos^2 \theta \rangle = 0.60$. Pump-probe angle $\alpha = 0°$. *Arrows* indicate the positions of minima. Adapted from [40]. © (2011) by the American Physical Society

HOMO shifts to the higher photon energy, but the interference with HOMO-2 shifts the minimum to even higher energies. In other words, the shift of the position of minimum in the HHG spectrum versus laser intensity cannot be attributed to a single factor alone.

6.6.2 Other Factors Influencing Precise Positions of HHG Minima

By using the angular distribution of molecules, the averaged PR transition dipole in Eq. (6.2) is calculated, thus it depends strongly on the alignment degree. Since the latter cannot be accurately measured in general, here I check how sensitive the calculated spectrum is with respect to the assumed alignment distribution. In Fig. 6.10c, it shows four different alignment distributions. They are multiplied by the volume element $\sin \theta$ for easy comparison.[11] Except for the commonly used $\cos^4 \theta$ distribution, the rest three of them are obtained from the calculated rotational wave packets,[12] with $\langle \cos^2 \theta \rangle$ as 0.63, 0.60 and 0.55, respectively. For 800-nm and 1200-nm lasers, the envelopes of HHG spectra resulting from different alignment distributions are shown in Fig. 6.10a and b, respectively. Except for the one from $\cos^4 \theta$ distribution, the precise position of minimum shifts slightly. However, it can be clearly seen that there is the change of a couple of eV's.

[11] The integral over ϕ' in Eq. (6.2) is incorporated into $\rho(\theta', \alpha)$, and $\theta = \theta'$ when $\alpha = 0°$ here.
[12] See Appendix B.

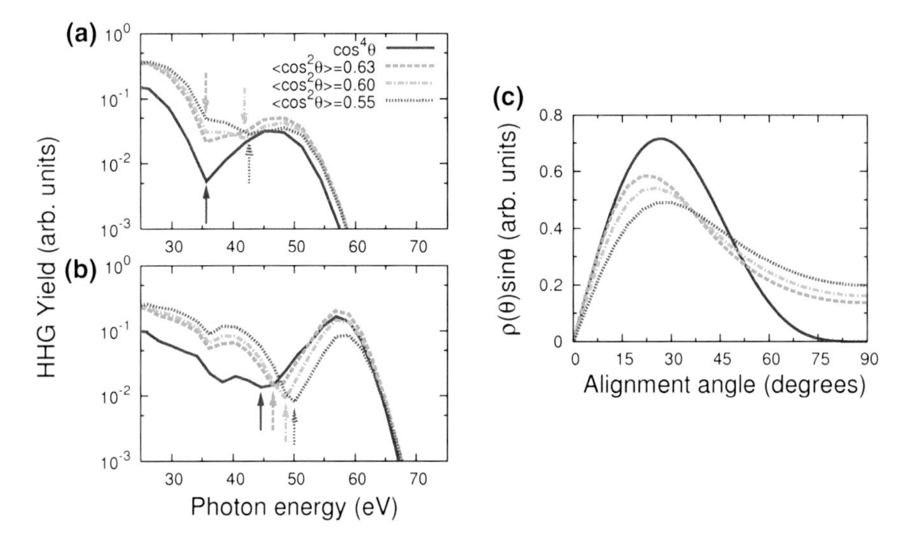

Fig. 6.10 Dependence of macroscopic HHG spectra (envelope only) with molecular alignment distributions for **a** an 800-nm laser with the intensity of 1.8×10^{14} W/cm^2, and **b** a 1200-nm laser with the intensity of 1.0×10^{14} W/cm^2. Weighted angular distributions of molecules are shown in **c**. Pump-probe angle $\alpha = 0°$. *Arrows* indicate positions of minima. Adapted from [40]. © (2011) by the American Physical Society

The accurate alignment-dependent ionization probability $N(\theta')$ for each molecular orbital is needed to precisely determine the minimum in the HHG spectrum. For CO_2 molecules, even for HOMO, non-negligible differences have been shown by different theories in the literature [2, 19, 23–28], and they didn't agree with the experimental data [23]. Here I examine how theoretical HHG spectra vary with different ionization rates used. For both the HOMO and the HOMO-2, ionization rates can all be easily calculated from the SFA or from the MO-ADK theory. Figure 6.11a and b show the simulated HHG spectra by using the ionization data shown in Fig. 6.11c and d. In figure captions other laser parameters used in the simulation are given. The shift of the position of minimum is 3 eV in Fig. 6.11a and 2 eV in Fig. 6.11b. Note that ionization probabilities from the SFA and the MO-ADK theory are normalized at the peak of the HOMO curve in Fig. 6.11c and d. The spectra are normalized at H33 (51 eV) in Fig. 6.11a and at H65 (67 eV) in Fig. 6.11b.

Due to the harmonic propagation in the medium and in free space, the measured HHG spectra are also sensitive to the experimental arrangement, and thus it probably can move the position of HHG minimum as well. I use another two geometries to demonstrate this argument. (i) Gas jet is put at the laser focus and HHG signal is collected using a slit; (ii) Gas jet is put after the laser focus, and HHG signal is collected without the slit (total signal). These two will be compared to the arrangement used in this thesis: putting the gas jet after the laser focus and collecting the HHG yield with a slit. In Fig. 6.12, the results are shown. Note that the spectra are normalized at H17 (26 eV) in Fig. 6.12a and at H35 (36 eV) in Fig. 6.12b. Both the spectra and the

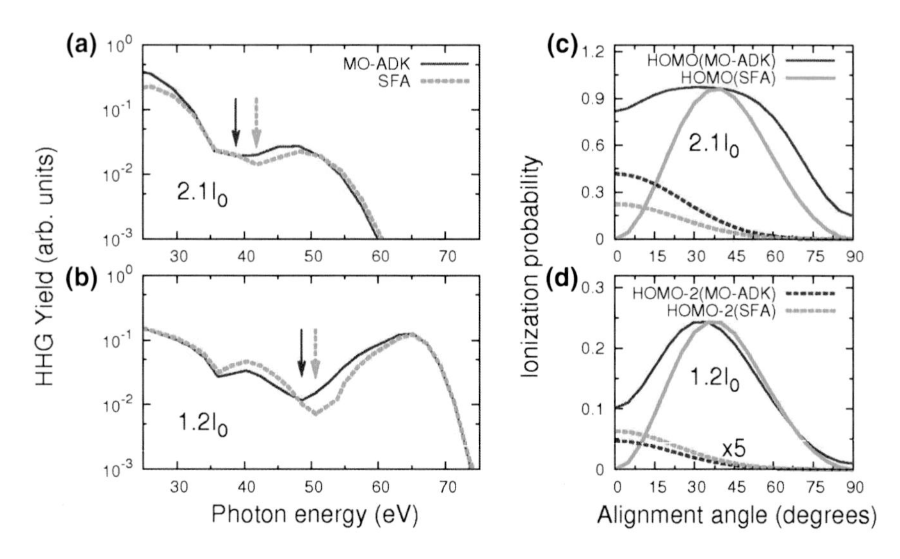

Fig. 6.11 Dependence of macroscopic HHG spectra (envelope only) on ionization probabilities calculated from the MO-ADK or the SFA in **a** an 800-nm laser, and **b** a 1200-nm laser. Laser intensities are indicated with $I_0 = 10^{14}$ W/cm^2. Alignment degree is $\langle \cos^2\theta \rangle = 0.60$. Pump-probe angle is $\alpha = 0°$. *Arrows* indicate the positions of minima. **c** and **d** Alignment-dependent ionization probabilities of HOMO and HOMO-2 calculated using the MO-ADK and the SFA. Ionization probabilities of HOMO-2 in **d** are multiplied by 5, and laser parameters are the same as **a** and **b**. Adapted from [40]. © (2011) by the American Physical Society

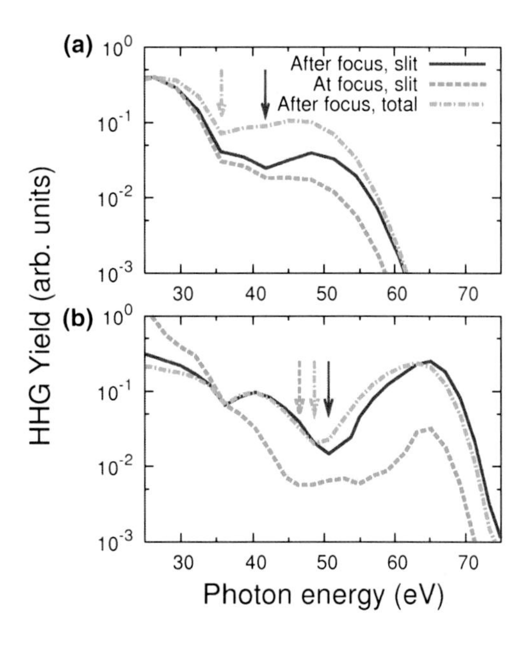

Fig. 6.12 Dependence of macroscopic HHG spectra (envelope only) on experimental arrangements **a** for an 800-nm laser with the intensity of 2.1×10^{14} W/cm^2, and **b** for a 1200-nm laser with the intensity of 1.2×10^{14} W/cm^2. Arrangements are: (1) gas jet is after laser focus and a slit is used (*solid lines*); (2) gas jet is at the focus and a slit is used (*dashed lines*); and (3) gas jet is after the focus but without the slit (*dot-dashed lines*). Alignment degree is $\langle \cos^2\theta \rangle = 0.60$. Pump-probe angle is $\alpha = 0°$. *Arrows* indicate positions of minima. Adapted from [40]. © (2011) by the American Physical Society

positions of HHG minima are changed quite significantly. In this comparison, the change of HHG spectrum is actually due to the change of MWP which depends on the experimental setup. Averaged PR transition dipoles of HOMO and HOMO-2 are the same in three calculations. These figures illustrate that to compare the position of HHG minimum from different experiments is very difficult.

6.7 Conclusion

In this chapter, I have calculated the macroscopic harmonic spectra of linear molecules in a gas medium by using a complete theory as discussed in Sect. 2.3.3. The approach is based on the simultaneous solutions of coupled Maxwell's wave equations describing the macroscopic propagation of both driving laser pulse and its high-harmonic fields together with the microscopic induced dipoles. For the latter I use the QRS theory. This scheme provides a simple and efficient method for calculating the HHG from a macroscopic medium. To my knowledge, this is the first time that theoretically calculated HHG spectra with including phase-matching and macroscopic propagation effects can be directly compared with experimentally measured spectra. I summarize this chapter in the following:

(1) I have demonstrated that experimental HHG spectra of random or aligned N_2 and CO_2 molecules can now be accurately reproduced by the theory. The contribution of outmost molecular orbital only or multiple molecular orbitals is included in the calculation depending on the conditions of molecular alignments. If there is only one molecular orbital dominant for the HHG process, I have further shown that HHG spectrum can be expressed as the product of a macroscopic wave packet and a photorecombination transition dipole (averaged over the alignment distribution). The latter is a property of the target, and is independent of HHG generating lasers. This factorization makes it possible to extract the target structure information from experimental HHG spectra. It provides the needed theoretical basis for using the HHG as ultrafast probes of excited molecules, such as those demonstrated in [3].
(2) I have carefully investigated HHG spectra of aligned N_2 molecules at a time delay of the maximal alignment when the pump-probe angle is set at 90° using a 1200-nm generating laser. At two low laser intensities, the minima in HHG spectra appear at about 38–40 eV. The minimum disappears at a higher laser intensity. To understand these results I have carried out the theoretical analysis and concluded that the minima in the HHG are associated with the properties of differential photoionization cross section from the HOMO. At higher intensity, the HOMO-1 contribution to the HHG becomes important and the interference between HOMO and HOMO-1 contributions washes out the minimum.
(3) I have examined the possibility of observing the well-known shape resonance in the HHG spectrum, which has been shown in the photoionization of N_2 both experimentally and theoretically. The normalized HHG yield (with respect to ran-

domly distributed molecules) at small alignment angles shows the clear enhancement due to the shape resonance. No evidence of the shape resonance has been observed in the HHG spectrum from 800-nm lasers, but it may have been seen in 1300-nm data [17]. It will be of great interest if further experiments can be dedicated to resolve this issue.

(4) I have analyzed the multiple orbital contribution to the HHG in aligned CO_2 molecules. Using lasers with different wavelengths and intensities there have been many experimental and theoretical studies on the CO_2 HHG from many laboratories, where CO_2 molecules are either randomly distributed or partially aligned. Particularly, many experiments have shown that HHG spectra exhibit minima and their positions shift with laser intensities [2, 38, 54] for CO_2 molecules that are partially aligned along the polarization axis of probe laser. The shift of minimum position with the laser intensity has been attributed to the interference between HOMO and HOMO-2. Both orbitals are contributing to the HHG, despite the HOMO-2 lying at 4.4 eV deeper than the HOMO. HHG spectra are quite sensitive to the experimental parameters, such as degree of alignment, laser focusing condition and use of a slit, however, the position of minimum in the HHG spectrum behaves in a similar trend as the laser intensity varies. This trend has been found to be consistent with the experimental measurements from different groups.

Comment that the QRS theory as shown in Sects. 2.2.3 and 2.3.3 entering into the macroscopic propagation equations is formulated in the time-independent fashion. The hole dynamics emphasized by Smirnova et al. [2] is not included in the QRS theory. Although the details between two theories are quite different, both explain the change of HHG minima with the laser intensity. In this chapter, it has been demonstrated the possibility of including phase-matching and macroscopic propagation effects "routinely" in the HHG theory for molecular targets. Further experimental investigations should explore the effects of laser focusing condition and gas pressure for lasers extending to longer wavelengths. Such studies would definitely further our basic understanding of the strong-field physics of molecules to the next level.

References

1. J. Itatani, J. Levesque, D. Zeidler, H. Niikura, H. Pépin, J.C. Kieffer, P.B. Corkum, D.M. Villeneuve, Tomographic imaging of molecular orbitals. Nature **432**, 867–871 (2004)
2. O. Smirnova, Y. Mairesse, S. Patchkovskii, N. Dudovich, D. Villeneuve, P. Corkum, MYu. Ivanov, High harmonic interferometry of multi-electron dynamics in molecules. Nature **460**, 972–977 (2009)
3. H.J. Wörner, J.B. Bertrand, D.V. Kartashov, P.B. Corkum, D.M. Villeneuve, Following a chemical reaction using high-harmonic interferometry. Nature **466**, 604–607 (2010)
4. S. Haessler, J. Caillat, P. Salières, Self-probing of molecules with high harmonic generation. J. Phys. B **44**, 203001 (2011)

5. A.T. Le, R.R. Lucchese, M.T. lee, and C.D. Lin. Probing molecular frame photoionization via laser generated high-order harmonics from aligned molecules. Phys. Rev. Lett. **102**, 203001 (2009)
6. H. Stapelfeldt, T. Seideman, Colloquium: Aligning molecules with strong laser pulses. Rev. Mod. Phys. **75**, 543–557 (2003)
7. E.W. Plummer, T. Gustafsson, W. Gudat, D.E. Eastman, Partial photoionization cross sections of N_2 and CO using synchrotron radiation. Phys. Rev. A **15**, 2339–2355 (1977)
8. A. Hamnett, W. Stoll, C.E. Brion, Photoelectron branching ratios and partial ionization cross-sections for CO and N_2 in the energy range 1850 e V. J. Electron Spectrosc. Relat. Phenom. **8**, 367–376 (1976)
9. R.R. Lucchese, G. Raseev, V. McKoy, Studies of differential and total photoionization cross sections of molecular nitrogen. Phys. Rev. A **25**, 2572–2587 (1982)
10. B.K. McFarland, J.P. Farrell, P.H. Bucksbaum, M. Gühr, High harmonic generation from multiple orbitals in N_2. Science **322**, 1232–1235 (2008)
11. Y. Mairesse, J. Higuet, N. Dudovich, D. Shafir, B. Fabre, E. Mével, E. Constant, S. Patchkovskii, Z. Walters, MYu. Ivanov, O. Smirnova, High harmonic spectroscopy of multichannel dynamics in strong-field ionization. Phys. Rev. Lett. **104**, 213601 (2010)
12. W. Boutu, S. Haessler, H. Merdji, P. Breger, G. Waters, M. Stankiewicz, L.J. Frasinski, R. Taieb, J. Caillat, A. Maquet, P. Monchicourt, B. Carre, P. Saliéres, Coherent control of attosecond emission from aligned molecules. Nat. Phys. **4**, 545–549 (2008)
13. T. Kanai, E.J. Takahashi, Y. Nabekawa, K. Midorikawa, Observing molecular structures by using high-order harmonic generation in mixed gases. Phys. Rev. A **77**, 041402 (2008)
14. N. Wagner, X. Zhou, R. Lock, W. Li, A. Wüest, M. Murnane, H. Kapteyn, Extracting the phase of high-order harmonic emission from a molecule using transient alignment in mixed samples. Phys. Rev. A **76**, 061403 (2007)
15. X. Zhou, R. Lock, W. Li, N. Wagner, M.M. Murnane, H.C. Kapteyn, Molecular recollision interferometry in high harmonic generation. Phys. Rev. Lett. **100**, 073902 (2008)
16. C. Vozzi, F. Calegari, E. Benedetti, J.-P. Caumes, G. Sansone, S. Stagira, M. Nisoli, R. Torres, E. Heesel, N. Kajumba, J.P. Marangos, C. Altucci, R. Velotta, Controlling two-center interference in molecular high harmonic generation. Phys. Rev. Lett. **95**, 153902 (2005)
17. R. Torres, T. Siegel, L. Brugnera, I. Procino, J.G. Underwood, C. Altucci, R. Velotta, E. Springate, C. Froud, I.C.E. Turcu, MYu. Ivanov, O. Smirnova, J.P. Marangos, Extension of high harmonic spectroscopy in molecules by a 1300 nm laser field. Opt. Express **18**, 3174–3180 (2010)
18. C.D. Lin, X.M. Tong, Z.X. Zhao, Effects of orbital symmetries on the ionization rates of aligned molecules by short intense laser pulses. J. Mod. Opt. **53**, 21–33 (2006)
19. S.-F. Zhao, C. Jin, A.T. Le, T.F. Jiang, C.D. Lin, Determination of structure parameters in strong-field tunneling ionization theory of molecules. Phys. Rev. A **81**, 033423 (2010)
20. X.M. Tong, Z.X. Zhao, C.D. Lin, Theory of molecular tunneling ionization. Phys. Rev. A **66**, 033402 (2002)
21. V.-H. Le, N.-T. Nguyen, C. Jin, A.T. Le, C.D. Lin, Retrieval of interatomic separations of molecules from laser-induced high-order harmonic spectra. J. Phys. B **41**, 085603 (2008)
22. C. Jin, J.B. Bertrand, R.R. Lucchese, H.J. Wörner, P.B. Corkum, D.M. Villeneuve, A.T. Le, C.D. Lin, Intensity dependence of multiple-orbital contributions and shape resonance in high-order harmonic generation of aligned N_2 molecules. Phys. Rev. A **85**, 013405 (2012)
23. D. Pavičić, K.F. Lee, D.M. Rayner, P.B. Corkum, D.M. Villeneuve, Direct measurement of the angular dependence of ionization for N_2, O_2, and CO_2 in intense laser fields. Phys. Rev. Lett. **98**, 243001 (2007)
24. M. Spanner, S. Patchkovskii, One-electron ionization of multielectron systems in strong non-resonant laser fields. Phys. Rev. A **80**, 063411 (2009)
25. S.-K. Son, S.-I. Chu, Multielectron effects on the orientation dependence and photoelectron angular distribution of multiphoton ionization of CO_2 in strong laser fields. Phys. Rev. A **80**, 011403 (2009)

26. M. Abu-samha, L.B. Madsen, Theory of strong-field ionization of aligned CO_2. Phys. Rev. A **80**, 023401 (2009)
27. S. Petretti, Y.V. Vanne, A. Saenz, A. Castro, P. Decleva, Alignment-dependent ionization of N_2, O_2, and CO_2 in intense laser fields. Phys. Rev. Lett. **104**, 223001 (2010)
28. S.-F. Zhao, C. Jin, A.T. Le, T.F. Jiang, C.D. Lin, Analysis of angular dependence of strong field tunneling ionization for CO_2. Phys. Rev. A **80**, 051402 (2009)
29. A.T. Le, R.R. Lucchese, C.D. Lin, Polarization and ellipticity of high-order harmonics from aligned molecules generated by linearly polarized intense laser pulses. Phys. Rev. A **82**, 023814 (2010)
30. A.T. Le, R.R. Lucchese, C.D. Lin, Uncovering multiple orbitals influence in high-harmonic generation from aligned N_2. J. Phys. B **42**, 211001 (2009)
31. X.X. Zhou, X.M. Tong, Z.X. Zhao, C.D. Lin, Role of molecular orbital symmetry on the alignment dependence of high-order harmonic generation with molecules. Phys. Rev. A **71**, 061801 (2005)
32. X.X. Zhou, X.M. Tong, Z.X. Zhao, C.D. Lin, Alignment dependence of high-order harmonic generation from N_2 and O_2 molecules in intense laser fields. Phys. Rev. A **72**, 033412 (2005)
33. M. Lein, N. Hay, R. Velotta, J.P. Marangos, P.L. Knight, Role of the intramolecular phase in high-harmonic generation. Phys. Rev. Lett. **88**, 183903 (2002)
34. C.D. Lin, A.T. Le, Z. Chen, T. Morishita, R. Lucchese, Strong-field rescattering physics-self-imaging of a molecule by its own electrons. J. Phys. B **43**, 122001 (2010)
35. A.T. Le, R.R. Lucchese, S. Tonzani, T. Morishita, C.D. Lin, Quantitative rescattering theory for high-order harmonic generation from molecules. Phys. Rev. A **80**, 013401 (2009)
36. T. Morishita, A.T. Le, Z. Chen, C.D. Lin, Accurate retrieval of structural information from laser-induced photoelectron and high-order harmonic spectra by few-cycle laser pulses. Phys. Rev. Lett. **100**, 013903 (2008)
37. G.H. Lee, I.J. Kim, S.B. Park, T.K. Kim, Y.S. Lee, C.H. Nam, Alignment dependence of high harmonics contributed from HOMO and HOMO-1 orbitals of N_2 molecules. J. Phys. B **43**, 205602 (2010)
38. H.J. Wörner, J.B. Bertrand, P. Hockett, P.B. Corkum, D.M. Villeneuve, Controlling the interference of multiple molecular orbitals in high-harmonic generation. Phys. Rev. Lett. **104**, 233904 (2010)
39. C. Jin, H.J. Wörner, V. Tosa, A.T. Le, J.B. Bertrand, R.R. Lucchese, P.B. Corkum, D.M. Villeneuve, C.D. Lin, Separation of target structure and medium propagation effects in high-harmonic generation. J. Phys. B **44**, 095601 (2011)
40. C. Jin, A.T. Le, C.D. Lin, Analysis of effects of macroscopic propagation and multiple molecular orbitals on the minimum in high-order harmonic generation of aligned CO_2. Phys. Rev. A **83**, 053409 (2011)
41. B.K. McFarland, J.P. Farrell, P.H. Bucksbaum, M. Gühr, High-order harmonic phase in molecular nitrogen. Phys. Rev. A **80**, 033412 (2009)
42. J.P. Farrell, B.K. McFarland, M. Gühr, P.H. Bucksbaum, Relation of high harmonic spectra to electronic structure in N_2. Chem. Phys. **366**, 15–21 (2009)
43. H.J. Wörner, H. Niikura, J.B. Bertrand, P.B. Corkum, D.M. Villeneuve, Observation of electronic structure minima in high-harmonic generation. Phys. Rev. Lett. **102**, 103901 (2009)
44. S. Minemoto, T. Umegaki, Y. Oguchi, T. Morishita, A.T. Le, S. Watanabe, H. Sakai, Retrieving photorecombination cross sections of atoms from high-order harmonic spectra. Phys. Rev. A **78**, 061402 (2008)
45. D.A. Telnov, S.-I. Chu, Effects of electron structure and multielectron dynamical response on strong-field multiphoton ionization of diatomic molecules with arbitrary orientation: An all-electron time-dependent density-functional-theory approach. Phys. Rev. A **79**, 041401 (2009)
46. V. Strelkov, Role of autoionizing state in resonant high-order harmonic generation and attosecond pulse production. Phys. Rev. Lett. **104**, 123901 (2010)
47. M. Tudorovskaya, M. Lein, High-order harmonic generation in the presence of a resonance. Phys. Rev. A **84**, 013430 (2011)

48. M.V. Frolov, N.L. Manakov, A.A. Silaev, N.V. Vvedenskii, A.F. Starace, High-order harmonic generation by atoms in a few-cycle laser pulse: Carrier-envelope phase and many-electron effects. Phys. Rev. A **83**, 021405 (2011)
49. C. Jin, A.T. Le, C.A. Trallero-Herrero, C.D. Lin, Generation of isolated attosecond pulses in the far field by spatial filtering with an intense few-cycle mid-infrared laser. Phys. Rev. A **84**, 043411 (2011)
50. A.D. Shiner, B.E. Schmidt, C. Trallero-Herrero, H.J. Wörner, S. Patchkovskii, P.B. Corkum, J.-C. Kieffer, F. Légaré, D.M. Villeneuve, Probing collective multi-electron dynamics in xenon with high-harmonic spectroscopy. Nature Phys. **7**, 464–467 (2011)
51. K. Kato, S. Minemoto, H. Sakai, Suppression of high-order-harmonic intensities observed in aligned CO_2 molecules with 1300-nm and 800-nm pulses. Phys. Rev. A **84**, 021403 (2011)
52. Y. Mairesse, J. Levesque, N. Dudovich, P.B. Corkum, D.M. Villeneuve, High harmonic generation from aligned moleculesamplitude and polarization. J. Mod. Opt. **55**, 2591–2602 (2008)
53. S. Haessler, J. Caillat, W. Boutu, C. Giovanetti-Teixeira, T. Ruchon, T. Auguste, Z. Diveki, P. Breger, A. Maquet, B. Carré, R. Taïeb, P. Salières, Attosecond imaging of molecular electronic wavepackets. Nature Phys. **6**, 200–206 (2010)
54. R. Torres, T. Siegel, L. Brugnera, I. Procino, J.G. Underwood, C. Altucci, R. Velotta, E. Springate, C. Froud, I.C.E. Turcu, S. Patchkovskii, MYu. Ivanov, O. Smirnova, J.P. Marangos, Revealing molecular structure and dynamics through high-order harmonic generation driven by mid- IR fields. Phys. Rev. A **81**, 051802 (2010)
55. O. Smirnova, S. Patchkovskii, Y. Mairesse, N. Dudovich, M. Yu Ivanov, Strong-field control and spectroscopy of attosecond electron-hole dynamics in molecules. Proc. Natl. Acad. Sci. USA **106**, 16556–16561 (2009)

Chapter 7
Photoelectron Angular Distributions in Single-Photon Ionization of Aligned N_2 and CO_2 Molecules Using XUV Light

7.1 Introduction

The photoionization (PI) is a basic physical process that allows one to directly investigate the molecular structure [1–4]. However, the rich dynamical structure of photoelectron angular distributions (PADs) for fixed-in-space molecules predicted about 30 years ago remains largely unexplored [5] because most of experimental measurements have been performed from an ensemble of randomly distributed molecules in the past. If the molecular cation dissociates immediately after the absorption of photon, the molecular frame photoelectron angular distribution (MF-PAD) can be investigated with X-ray or extreme ultraviolet (XUV) photons, while the molecular axis can be inferred from the direction of the motion of fragmentation products by using photoion-photoelectron coincidence techniques. This method clearly cannot be applicable if molecular cations are stable, thus it is not applicable to the PI from the highest-occupied molecular-orbital (HOMO). It has been shown in the last decade that gas-phase molecules can be transitionally aligned with infrared (IR) lasers by using either adiabatic or nonadiabatic methods [6–8]. It is preferable to investigate the PI of molecules by using the nonadiabatic method since the molecules can be aligned in field-free conditions.

In 2008, Thomann et al. [9] measured the angular dependence of single-photon ionization of aligned N_2 and CO_2 molecules by using the high-order harmonic generation (HHG) light. They used a non-ionizing IR laser pulse impinging on supersonically cooled molecules, and then molecules were impulsively aligned. These molecules were ionized by 10-fs XUV pulses, and resulting photoelectrons were collected in coincidence with the molecular ions. The rotational wave packet generated by aligning pulse periodically exhibits the macroscopic field-free alignment with respect to the polarization axis of pump laser. The XUV light was from the HHG. Only 27th harmonic (H27) ($\sim$43 eV) was selected by an Aluminum filter. Both ions and photoelectrons were measured during the first half revival where the molecular distribution changes from aligned to antialigned. In other words, the molecular-frame PI can be directly probed by investigating the PI from partially aligned molecules.

C. Jin, *Theory of Nonlinear Propagation of High Harmonics Generated in a Gaseous Medium*, Springer Theses, DOI: 10.1007/978-3-319-01625-2_7,

However, Thomann et al.'s [9] measurement did not report the PAD from aligned molecules due to the limited number of XUV photons. Similarly, different ionic states cannot be distinguished because of the energy resolution. Despite this, it was clearly established in this experiment that the PI yield was maximum when molecules were aligned perpendicular to the polarization of XUV light for both N_2 and CO_2 molecules. Note that single-photon ionization here is in contrast to multiphoton (or tunneling) ionization in the strong field where two molecules show great differences.

In this chapter I will first explain Thomann et al.'s [9] experimental data by using the well established PI theory of molecules [10, 11]. The PI theory has been widely used to interpret the measurements for randomly distributed molecules, however, it also predicts the detailed molecular structures that can be observed from aligned molecules, in particular, the angular distributions of photoelectrons from different orbitals. It is expected that the PAD from aligned molecules become available soon due to the rapidly advancing technology of the HHG.[1] This chapter is organized as follows. In Sect. 7.2, I will examine the alignment dependence of H27 when it is generated by exposing N_2 molecules to intense IR lasers [13]. Because there is a recombination step involved in the HHG, which is the inverse process of photoionization, such a comparison is of interest. In Sect. 7.3, I will show the simulated single-photon ionization yield of aligned N_2 and CO_2 molecules in the experiment of Thomann et al. [9]. This would also allow one to compare the alignment dependence of the single-photon PI by XUV light with the multiphoton ionization by IR lasers [14, 15]. In Sect. 7.4, I will report the predicted PAD of fixed-in-space N_2 molecules in the laboratory frame. In Sect. 7.5, I will introduce the pump-probe setup similar to Thomann et al. [9] and report the predicted PAD for partially aligned N_2 molecules (with the different degree of alignment) in the laboratory frame. Predicted angular distributions in the laboratory frame are presented in such a way that they can be compared directly to future experiments. In fact, several such measurements have been reported recently [16, 17]. In Sect. 7.6, I will investigate how the PADs of aligned N_2 molecules change with the photon energy. In Sect. 7.7, I will predict the PAD in the laboratory frame for fixed-in-space and partially aligned CO_2 molecules by using the same pump-probe scheme as in Sect. 7.5. A summary in Sect. 7.8 will conclude this chapter.

7.2 Connection Between Photoionization and HHG

For aligned molecules, one can measure the ionization cross section experimentally by a single photon, say, at 43 eV, or by the multiphotons with IR lasers, say, at 800 nm (1.55 eV per photon), and measure the high-order harmonics, say, H27 ($\sim$43 eV). Experiments can be carried out at intense IR laser facilities because the field-free molecules are transiently aligned by IR lasers only. To compare with the experimental measurements, the theory must be performed for fixed-in-space molecules first. Here

[1] Kelkensberg et al. [12] have reported the PAD of aligned CO_2 molecules in 2011.

the photoionization of N_2 by a 43-eV photon leaves N_2^+ in $X\ ^2\Sigma_g^+$, $A\ ^2\Pi_u$ and $B\ ^2\Sigma_u^+$ ionic states, corresponding to removing an electron from HOMO, HOMO-1 and HOMO-2 orbitals of N_2, respectively. In Fig. 7.1a, the alignment dependence of integrated PICS for the ionization leading to X, A and B states in the ion are shown. The molecular axis makes an angle of θ with respect to the polarization axis of XUV light. According to Eq. (C.19), one can obtain the asymmetry parameter $\beta_{\hat{n}}$ to be -0.83, -0.95 and 0.47 for X, A and B states, respectively [10].

Note that PICSs are of the same order of magnitude in Fig. 7.1a. Even though the HOMO is a σ_g orbital and the HOMO-1 is a π_u orbital, two channels have nearly identical θ-dependence. The ionization by intense IR lasers is in the strong contrast to these behaviors. Fig. 7.1c shows the alignment-dependent multiphoton (or tunneling) ionization rates of N_2 molecules calculated with the MO-ADK theory [19, 20]. An intense IR laser with the peak intensity of 2×10^{14} W/cm^2 is applied. Since the tunneling is the main mechanism for the ionization at such high intensities, ionization rates for X, A and B states are quite different, and they decrease rapidly as the ionization energy increases. Note that A and B states are higher than X state (15.6 eV) by 1.3 and 3.2 eV, respectively [21]. It is known that the shape of the charge

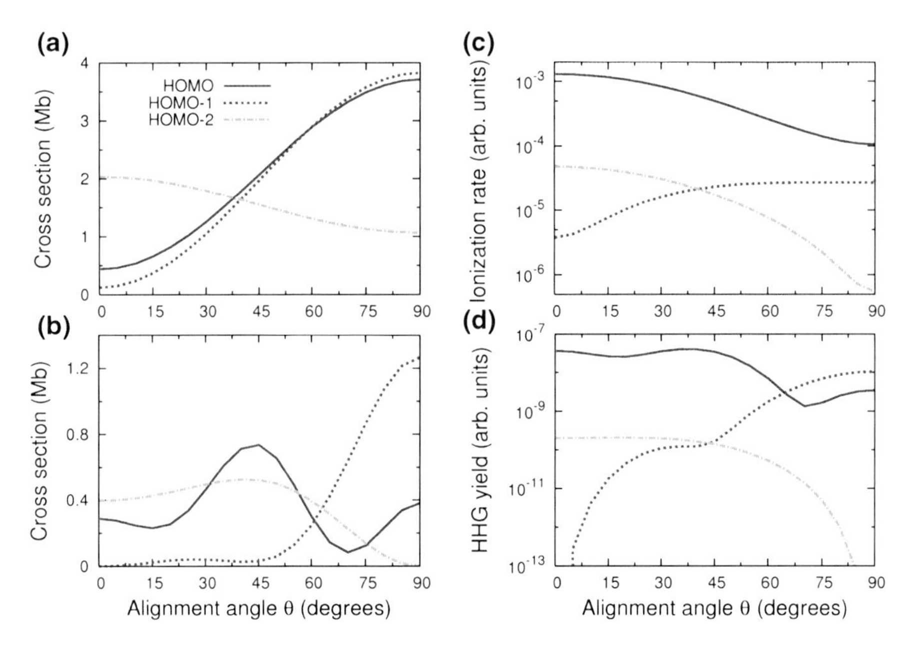

Fig. 7.1 **a** Integrated photoionization cross sections, and **b** doubly differential cross sections where electrons are ejected in the polarization direction, for N_2 aligned at an angle θ from the polarization axis, by a single photon at 43 eV. **c** Alignment dependence of multiphoton ionization rate of N_2 by a laser with the intensity of 2×10^{14} W/cm^2. **d** Alignment dependence of 27th harmonic ($\sim$43 eV) of N_2 by a laser with wavelength of 800 nm, intensity of 2×10^{14} W/cm^2 and duration (FWHM) of 30 fs. Three channels of HOMO (*solid lines*), HOMO-1 (*dotted lines*) and HOMO-2 (*dot dashed lines*) are plotted. Adapted from [18]. © (2010) by the American Physical Society

density of molecular orbital (from which the electron is removed) is also reflected by the alignment dependence of strong-field ionization.

Next I consider the alignment dependence of single-molecule HHG. From the quantitative rescattering (QRS) theory in Eq. (2.36), one knows that the HHG yield is proportional to the product of the differential PICS for electrons ejected along the polarization axis, see Fig. 7.1b, and the tunneling ionization rate, see Fig. 7.1c. Note that the modulus square of photoionization transition dipole is proportional to the differential PICS. In Fig. 7.1d, it shows the resulting alignment dependence of H27. When molecules are aligned near 90°, the HOMO-1 overtakes the HOMO in contributing to H27 at the intensity of 2×10^{14} W/cm^2. Le et al. [13] have explained these results observed experimentally by McFarland et al. [22] using the QRS theory. HHG contributions from three orbitals shown in Fig. 7.1d should have been added coherently. As discussed in Sects. 6.2 and 6.3, the coherence can be neglected when there is only one dominant channel.

One can retrieve the phase of photoionization transition dipole from the phase of high harmonic, see [23–26]. The former usually cannot be provided by the traditional PI measurements, however, the latter can be measured by the RABITT technique. On the other hand, unlike the photoionization, experimental HHG spectra suffer from phase-matching and macroscopic propagation effects as discussed in Chap. 6, thus the caution must be taken if one draws the conclusion on single molecule HHG spectra from experimental HHG spectra.

7.3 Total Photoionization Yield from Aligned N_2 and CO_2

Thomann et al. [9] employed the ultrafast high-harmonic XUV light to ionize field-free aligned molecules where cations are nondissociative. They selected a single harmonic order—H27 (~43 eV)—with a width of a few eV's as the probe ionizing pulse. The ion yield was detected around the first half-revival of 4.2 ps for N_2 and 21.0 ps for CO_2, respectively. Since the polarization axes of aligning and ionizing pulses were parallel, molecular distributions had the cylindrical symmetry with respect to the polarization axis. Yields of singly ionized N_2 and CO_2 by 43 eV photons were presented as a function of time delay between two pulses. In this section, I will compare theoretical calculations with their experimental results. The formulation for the alignment dependence of integrated PICS is presented in Appendix C.3.

7.3.1 Single-Photon Ionization Yield of Aligned N_2: Theory Versus Experiment

The calculated degree of alignment $\langle \cos^2\theta \rangle$ for N_2 versus time delay is shown in Fig. 7.2a. In the calculation, pump (or aligning) laser has the intensity of 5×10^{12}

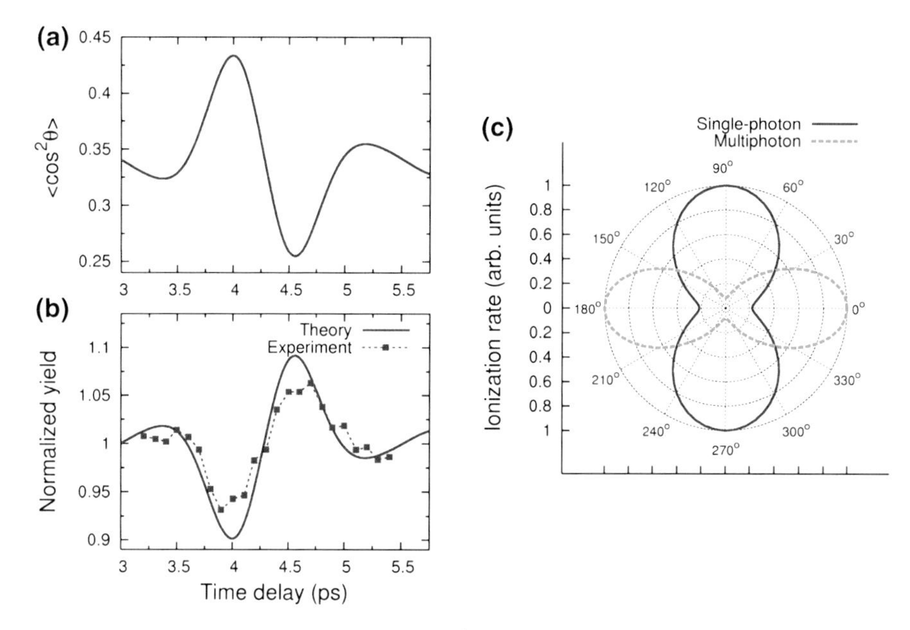

Fig. 7.2 **a** Calculated degree of alignment $\langle\cos^2\theta\rangle$ for N_2 versus time delay near the first half-revival. **b** Single-photon ionization yield from transiently aligned N_2 by 43 eV photons versus time delay: theory (*solid line*) and experiment (*solid squares*) [9]. **c** Angular dependence of the ionization rate by single-photon (43 eV) ionization (*solid line*), and by the multiphoton ionization using an IR laser with the intensity of 2×10^{14} W/cm^2 (*dashed line*). Adapted from [18]. © (2010) by the American Physical Society

W/cm^2, the wavelength of 800 nm and the duration (full width at half maximum, FWHM) of 140 fs, taken from [9]. Gas temperature is chosen as 20 K to obtain a high degree of alignment. $\langle\cos^2\theta\rangle$ can be calculated from Eq. (B.4) by solving the time-dependent Schrödinger equation of Eq. (B.1). In Fig. 7.2c, it shows the alignment-dependent PICS (normalized at $\theta = 90°$, and summed over X, A and B ionic states) of N_2. With the calculated alignment distribution of molecules, one can obtain the photoionization yield versus time delay. In Fig. 7.2b, the calculated results are shown for N_2 and compared to the experimental data in [9]. Clearly the theory and the experiment agree well. Both show that the total ionization yield peaks when the molecules are aligned mostly perpendicular to the polarization axis. This can be easily understood from Fig. 7.1a and from Fig. 7.2c with the combination of the alignment distribution and alignment-dependent PICS. For comparison, in Fig. 7.2c, it also shows the alignment dependence of multiphoton ionization rates (normalized at $\theta = 0°$) by intense IR lasers at the peak intensity of 2×10^{14} W/cm^2. The MO-ADK theory [19, 20] is used here to calculate the ionization rate. In this case, the ionization occurs mostly from the HOMO orbital, i.e., only from the X ionic state.

7.3.2 Single-Photon Ionization Yield of Aligned CO_2: Theory Versus Eexperiment

I carry out the similar calculations for CO_2 molecules. Parameters for the pump laser are intensity of 3.5×10^{12} W/cm^2, wavelength of 800 nm and duration (FWHM) of 140 fs, taken from [9]. Gas temperature is chosen to be 20 K. The photoionization of CO_2 by XUV light leads to CO_2^+ in X $^2\Pi_g$, A $^2\Pi_u$ and B $^2\Sigma_u^+$ ionic states, corresponding to removing an electron from HOMO, HOMO-1 and HOMO-2 orbitals of CO_2. $\langle\cos^2\theta\rangle$ and the total ionization yield versus time delay for CO_2 are shown in Fig. 7.3a and b, respectively. Both have behaviors that are quite similar to those seen in Fig. 7.2a and b for N_2. Calculated results in Fig. 7.3b are in good agreement with the results in [9]. In Fig. 7.3c, the integrated PICSs at each fixed alignment angle θ for the ionization leading to X, A and B ionic states are shown, similar to Fig. 7.1a. The asymmetry parameter $\beta_{\hat{n}}$ are -0.64, -0.77 and -0.53 for three ionic states.

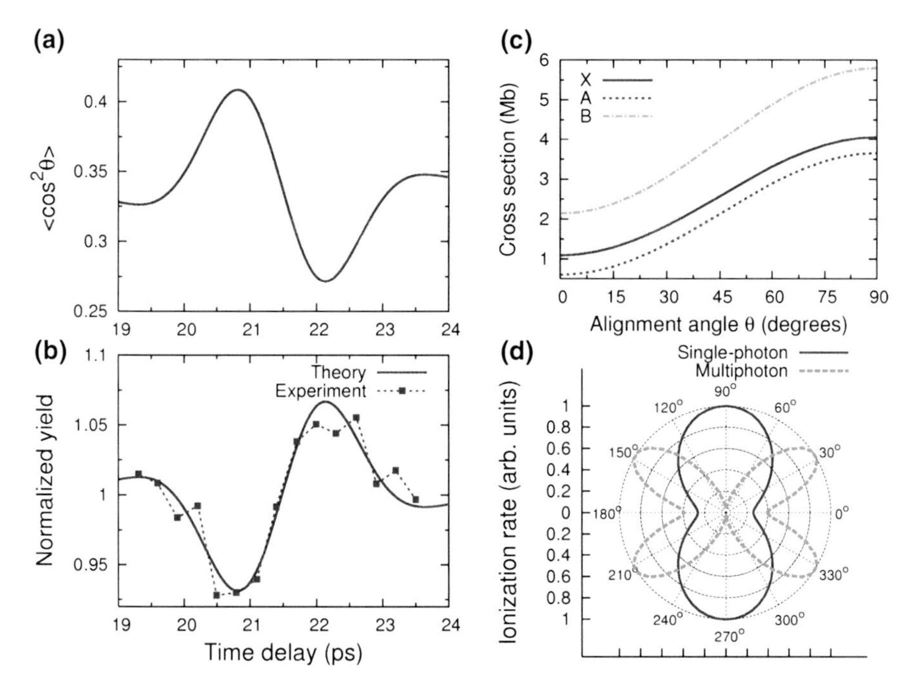

Fig. 7.3 **a** Calculated degree of alignment $\langle\cos^2\theta\rangle$ for CO_2 versus time delay near the first half-revival. **b** Single-photon ionization yield from transiently aligned CO_2 by 43 eV photons versus time delay: theory (*solid line*) and experiment (*solid squares*) [9]. **c** Integrated photoionization cross section for the ionization leading to X (*solid line*), A (*dotted line*) and B (*dot dashed line*) ionic states of CO_2^+, with the alignment angle θ, by the single-photon (43 eV) ionization of CO_2. **d** Angular dependence of the ionization rate for single-photon (43 eV) ionization (*solid line*) and multiphoton ionization by an infrared laser with the intensity of 1.1×10^{14} W/cm^2 (*dashed line*). Adapted from [18]. © (2010) by the American Physical Society

In Fig. 7.3d, the θ-dependence of ionization cross sections (summed over X, A and B states, and normalized at $\theta = 90°$) are shown in the polar plot. The total multiphoton ionization rate (normalized at $\theta = 35°$) versus θ for an IR laser with the intensity of 1.1×10^{14} W/cm^2 is also shown in Fig. 7.3d. The angular dependence of π_g orbital (HOMO) is reflected by the shape of a butterfly in the multiphoton ionization. Note that the degeneracy of molecular orbitals should be included in adding up the cross sections from different channels, and the alignment dependence of the multiphoton ionization for HOMO, HOMO-1 and HOMO-2 are actually quite different due to different oribtal symmetries. To obtain the tunneling ionization rate from the MO-ADK theory as shown in Fig. 7.3d, the vertical ionization energies are taken from [11, 21] while the molecular parameters are from Zhao et al. [20]. Similar alignment-dependent multiphoton ionization rates for different laser intensities have been shown in Fig. 6.11 by using the MO-ADK theory or the strong-field approximation.

7.4 Photoelectron Angular Distributions (PADs) of Fixed-in-Space N_2 in the Laboratory Frame

Experimentally, one can use the COLTRIMS[2] technique to measure the full momentum vectors of charged particles resulting from the ionization of molecules [27]. The detected photoelectron energy E_{pe} is related to the photon energy $h\nu$ and the vertical ionization energy E_{ion} by $E_{pe} = h\nu - E_{ion}$, where E_{ion} is for different ionization channels. So one can measure the PAD from different ionization channels. Here the results are presented in the laboratory frame so that they can be compared to future measurements. The formulation for the PAD in the laboratory frame is given in Appendix C.4.

In Fig. 7.4, the PADs from fixed-in-space N_2 molecules that make an angle θ with the polarization axis are shown. The 43-eV photon is used and ions are left in X, A and B ionic states after the PI. In Fig. 7.4a–c the PADs for emission angles from 0 to 90° at several molecular alignments are compared. The PAD changes rapidly as the alignment angle varies. For three channels, the PADs at a given molecular alignment angle also vary significantly. To compare the complicated PAD easily, color coding is used to present the PAD for each ionic state, see Fig. 7.4d–g. Photoelectron energies are measured by the radius. Thus the rings, starting from the outermost rings, are for electrons ejected from HOMO, HOMO-1 and HOMO-2, respectively. These figures show the complicated variations of PADs from aligned molecules. In contrast, the PAD with PI from isotropically distributed molecules depends on a single β-parameter, as expressed in Eq. (C.25).

[2] Cold Target Recoil Ion Momentum Spectroscopy.

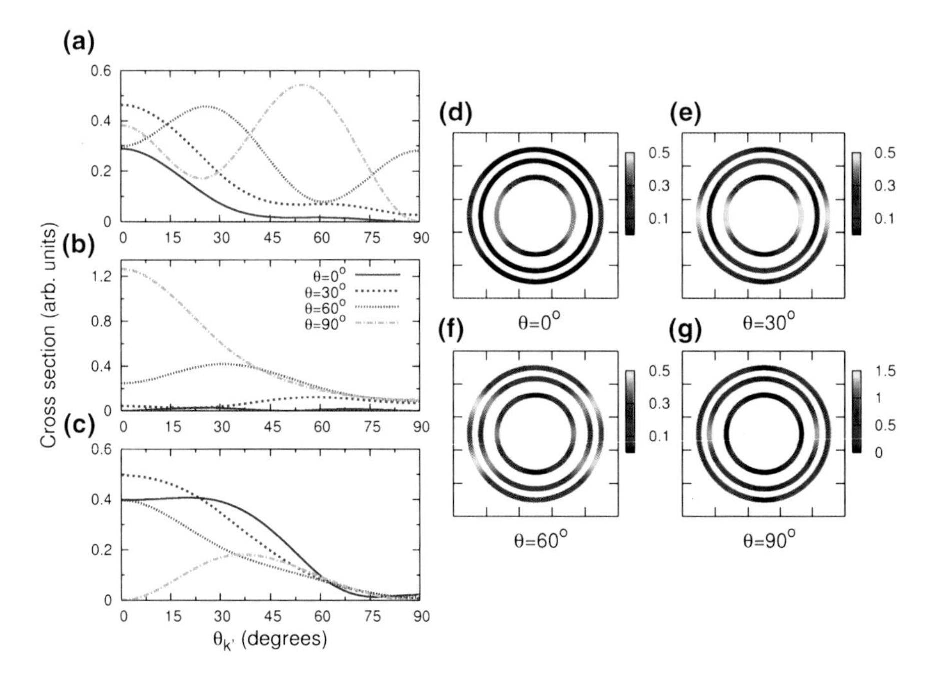

Fig. 7.4 Photoionization cross sections in the laboratory frame for the single-photon (43 eV) ionization of fixed-in-space N_2 versus the emission angle $\theta_{k'}$ at alignment angles indicated and for the ionization leading to N_2^+ in X, A and B states, shown in panels (**a**–**c**), respectively. In panels (**d**–**g**) the same distributions are shown for X, A and B channels at each fixed-in-space molecular alignment angle. See text. Adapted from [18].

7.5 PADs of Transiently Aligned N_2 in the Laboratory Frame

7.5.1 PADs at Low Degree of Alignment

For transiently aligned molecules the angular distribution with respect to the laser polarization axis evolves with time delay. The PAD for each fixed emission angle must be averaged over the molecular alignment distributions to compare with the experimental measurements. In Fig. 7.5a–c, the PADs after such averaging for X, A and B ionic states, are shown at two time delays when molecules are either maximally aligned or antialigned. These plots clearly show that angular averaging has severely smoothed out the structures in comparison with "raw" data as shown in Fig. 7.4. In Fig. 7.5a–c, the PADs for isotropically distributed molecules are also shown for comparison. One can find the values of $\beta_{\hat{k}'}$ to be 0.74, 1.20 and 1.88 for X, A and B channels, respectively, by using Eq. (C.25) [10]. The PADs from three ionic states are compared together in Fig. 7.5d and e using the color coding. Since the PAD has been expressed in the laboratory frame with the fixed polarization axis, such data can

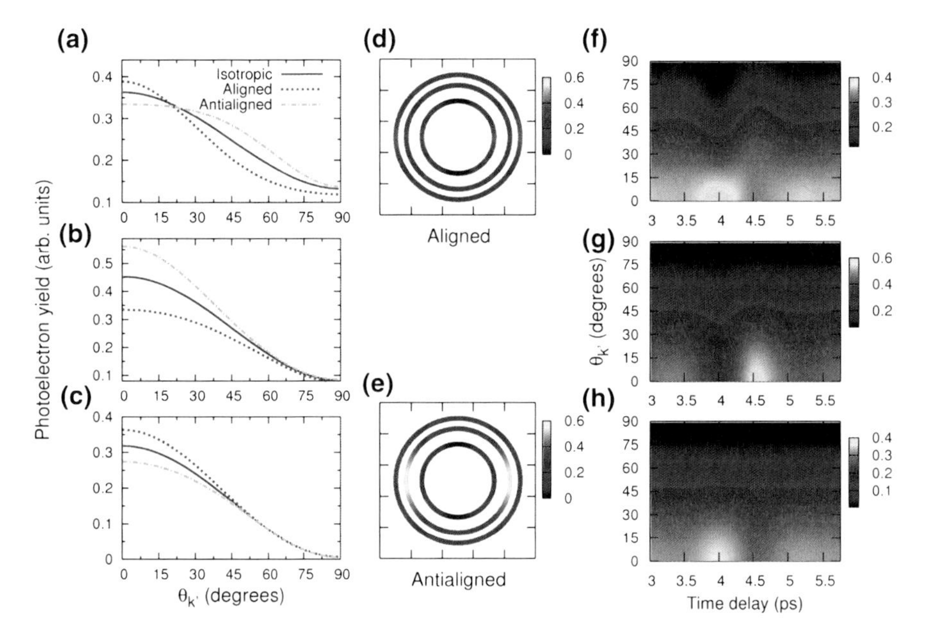

Fig. 7.5 PADs in the laboratory frame for single-photon (43 eV) ionization of N_2 as a function of emission angle $\theta_{k'}$ and pump-probe time delay. **a–c**: Molecules are maximally aligned ($\tau = 4.00$ ps), antialigned ($\tau = 4.55$ ps) and isotropically distributed, for the ionization leading to N_2^+ ions in X, A and B states, respectively. **d** and **e**: The same distributions are compared for maximally aligned and antialigned molecules. **f–h**: PADs versus time delay for X, A and B channels, respectively. Adapted from [18]. © (2010) by the American Physical Society

be compared directly to future experiments. Alternatively, one can also measure the laboratory-fixed PAD for each ionic state versus time delay, and they are shown in Fig. 7.5f–h. In the future the experimental data like Fig. 7.5f–h can be de-convoluted to retrieve the PAD for fixed molecular alignment angles, and then compared with the theoretical calculations as shown in Fig. 7.4a–c.

7.5.2 PADs at High Degree of Alignment

In Fig. 7.5, the pump laser used to align molecules is the same as that assumed in Fig. 7.2, the maximum degree of alignment is only $\langle \cos^2\theta \rangle = 0.43$. Thus there are no striking features observed. To improve the contrast, I assume a pump laser with the intensity of 5×10^{13} W/cm^2 and the duration of 60 fs at the temperature of 20 K to align molecules. The maximum degree of alignment achieved is $\langle \cos^2\theta \rangle = 0.71$. In this case, the molecules are maximally aligned at $\tau = 4.04$ ps and antialigned at $\tau = 4.39$ ps. The PADs in the laboratory frame are shown in Fig. 7.6. With the better alignment, one can clearly see the improved contrast in comparison with Fig. 7.5.

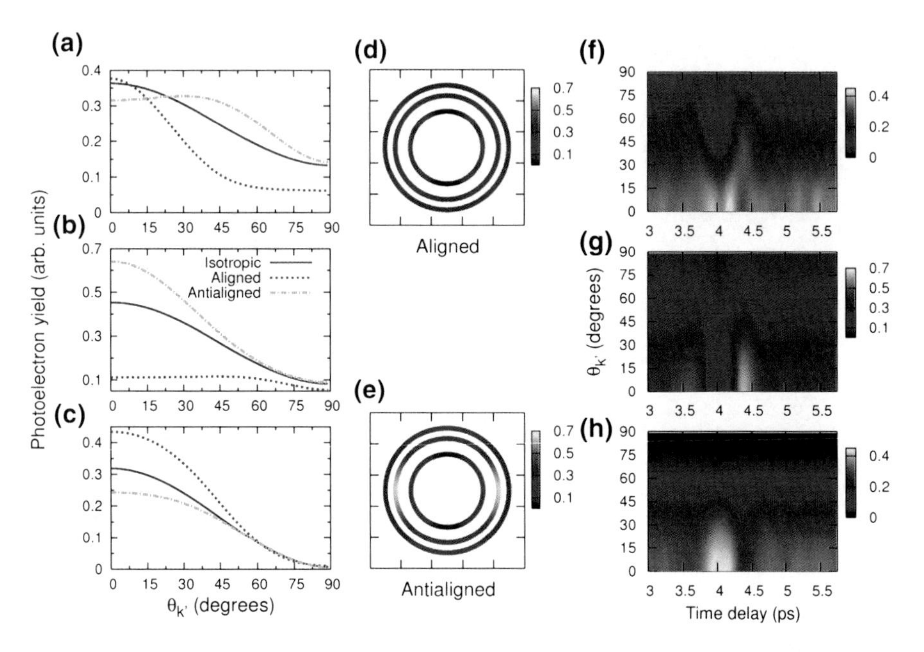

Fig. 7.6 Same as Fig. 7.5 except that a strong aligning pump laser is assumed. See text. Adapted from [18]. © (2010) by the American Physical Society

And the PADs of aligned and antialigned molecules become closer to that for fixed-in-space molecules in Fig. 7.5 at $\theta = 0$ and $90°$, respectively.

7.6 Photon Energy Dependence of PADs for Aligned N_2

All calculations above have been performed at the photon energy of 43 eV. Next the behavior of alignment dependence of the PAD at other photon energies is explored. For example, the photon energy is at 20, 30 or 46 eV. In Fig. 7.7, it shows the PADs for X, A and B channels at the fixed alignment angles of $\theta = 0$, 30, 60 and $90°$, together with the PADs for isotropically distributed molecules. The PAD varies substantially at each alignment angle as the photon energy changes. There is a $3\sigma_g \rightarrow k\sigma_u$ resonance in the HOMO channel at 30 eV, which is known as the shape resonance, and this resonance occurs for small alignment angles only in Fig. 7.7e.[3] Figure 7.7 clearly shows that the PAD for aligned molecules is quite complicated, and the PAD changes much with the photon energy as well as with the alignment angle. Thus one

[3] The same conclusion is obtained in Fig. 6.5, where the PAD of N_2 at $\theta_{k'} = 0°$ is shown.

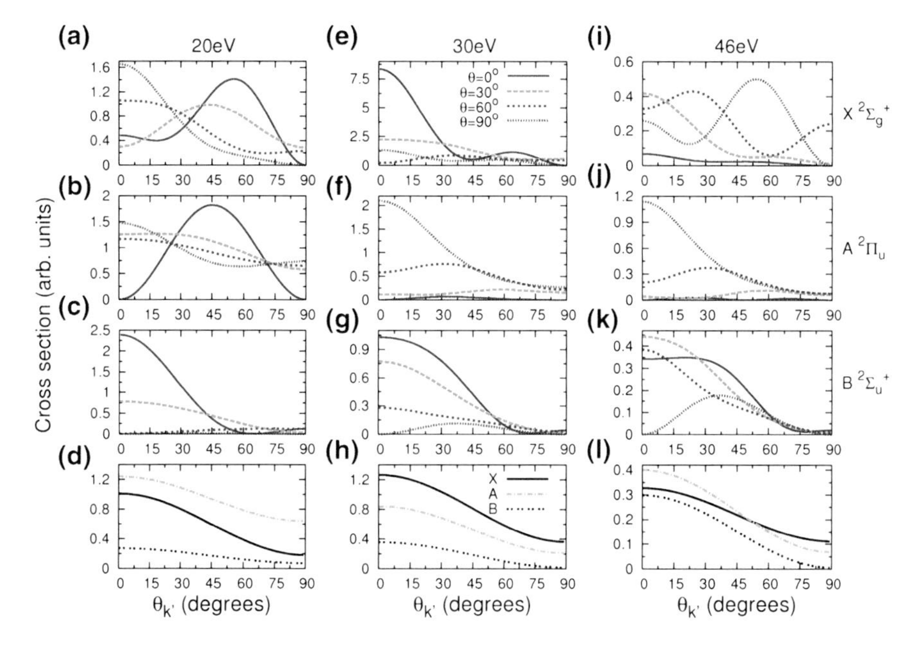

Fig. 7.7 Fixed-in-space photoionization angular distributions (versus $\theta_{k'}$) in the laboratory frame, for the photon energies of 20, 30 or 46 eV, and four alignment angles are shown. *First row*: X-channel. *Second row*: A-channel. *Third row*: B-channel. The *last row* shows that the PAD becomes featureless if the molecules are isotropically distributed. Adapted from [18]. © (2010) by the American Physical Society

can conclude that the measurements that do not explore the alignment dependence will tend to miss the important features in trying to understand the dynamics of a molecule.

7.7 PADs of Transiently Aligned CO_2 in the Laboratory Frame

7.7.1 PADs of Fixed-in-space CO_2

In Fig. 7.8, the PADs in the laboratory frame for CO_2 molecules ionized by a 43 eV photon are shown. Three ionization channels of $X\ ^2\Pi_g$, $A\ ^2\Pi_u$ and $B\ ^2\Sigma_u^+$ have the ionization potential of 13.8, 17.7 and 18.2 eV, respectively [11, 21]. The PADs are shown at alignment angles $\theta = 0$, 30, 60 and 90° in Fig. 7.8a–c. Using false colors, the PADs from three channels are compared together in Fig. 7.8d–g. Again the energy of photoelectron is measured by the radius of circle.

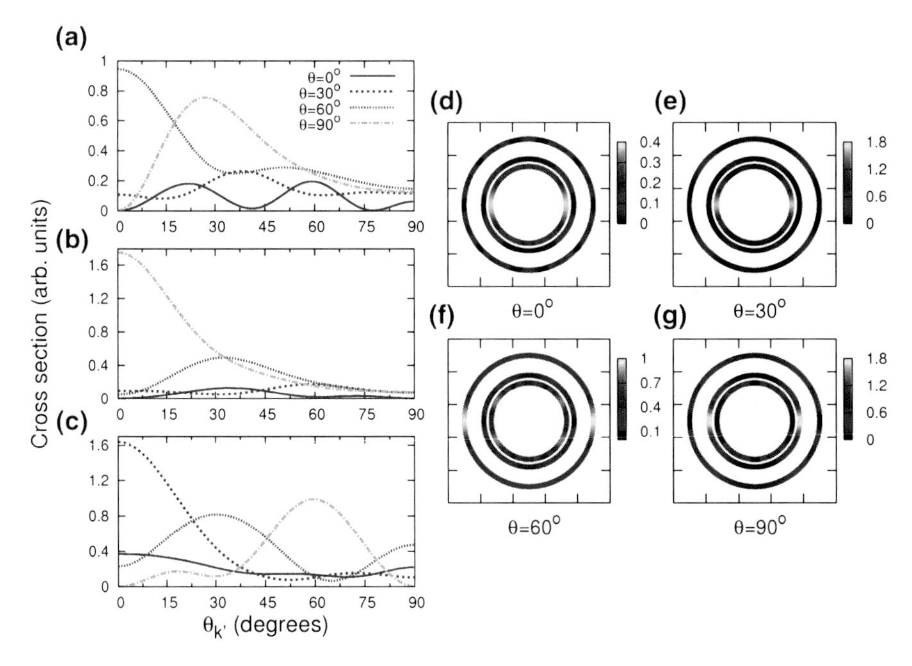

Fig. 7.8 Photoionization cross sections in the laboratory frame for single-photon (43 eV) ionization of fixed-in-space CO_2 versus emission angle $\theta_{k'}$ at alignment angles indicated and for the ionization leading to CO_2^+, in panels (**a**–**c**), in X, A and B states, respectively. In panels (**d**–**g**) the same distributions are shown for X, A and B channels at each fixed-in-space molecular alignment angle. See text. Adapted from [18]. © (2010) by the American Physical Society

7.7.2 PADs of Aligned CO_2

In Fig. 7.9a–c, I show the calculated PADs for CO_2 molecules aligned by a pump laser, as well as when the molecules are isotropically distributed. Molecules are aligned and antialigned at the time delay of $\tau = 20.82$ ps and 22.14 ps, respectively. In Fig. 7.9d and e, the PADs are compared together, and in Fig. 7.9f–h, they are expressed as a function of time delay. One can deduce that the asymmetry parameters, $\beta_{\hat{k}'}$, in Eq. (C.25), are 0.92, 1.32 and 0.68 for X, A and B states, respectively [11], in Fig. 7.9a–c. In the future when the experimental data similar to Fig. 7.9f–h become available, one may de-convolute the experimental results to retrieve the alignment dependence of PAD and compare to the calculated values assuming the angular distribution of molecules is known.

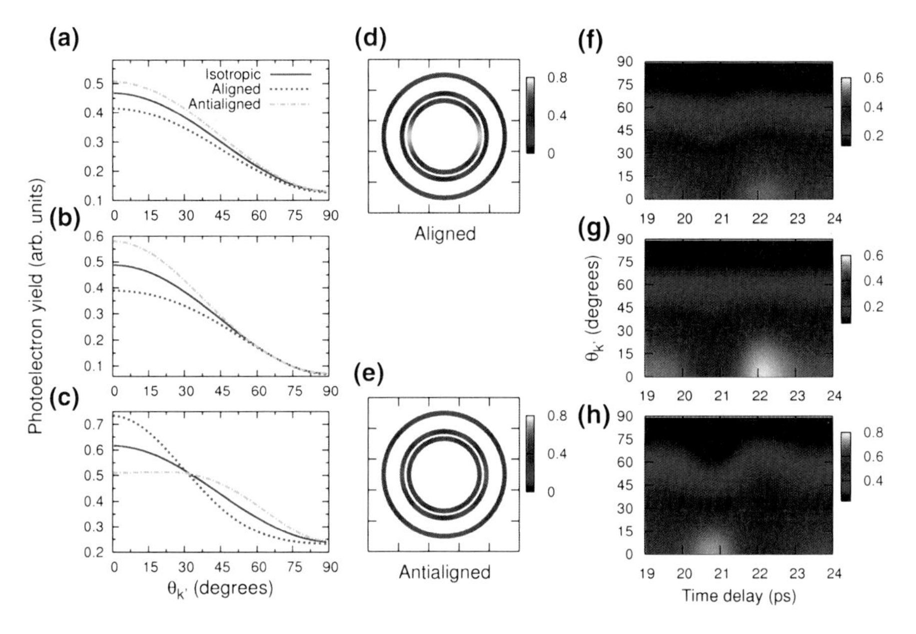

Fig. 7.9 PADs in the laboratory frame for single-photon (43 eV) ionization of CO_2 as a function of emission angle $\theta_{k'}$ and pump-probe time delay. **a–c**: Molecules are maximally aligned ($\tau = 20.82$ ps), antialigned ($\tau = 22.14$ ps) and isotropically distributed, for the ionization leading to CO_2^+ ions in X, A and B states, respectively. **d** and **e**: The same distributions are compared for maximally aligned and antialigned molecules. **f–h**: PADs versus time delay for X, A and B channels, respectively. Adapted from [18]. © (2010) by the American Physical Society

7.8 Conclusion

In this chapter I have studied the photoelectron angular distributions (PADs) from aligned N_2 and CO_2 molecules. Compared to the PADs from isotropically distributed molecules [5], these data can provide much more details on the molecule. By using an IR laser the field-free molecular alignment can be achieved, so one can measure the PAD from aligned molecules at the strong-field IR laser facilities only. One can use IR laser to align molecules, to generate soft X-ray or XUV photons and to ionize aligned molecules. Experiments dedicated to measure the total ionization yield from such aligned molecules have been carried out. With the higher intensity of XUV photons or soft X-rays becoming available, one can measure the PAD from aligned molecules. In fact, such measurement has been reported recently [12]. For geometries where the PAD can be measured in the future I have provided the theoretical predictions. Using aligned linear molecules such as N_2 and CO_2, I have calculated the expected PAD for the electron removal from HOMO, HOMO-1 and HOMO-2 orbitals, by the well-established photoionization codes. These theoretical results often have not been tested except for randomly distributed molecules. In the future, the PAD from aligned molecules also undergoing changes in vibrational levels can be measured.

In fact, such experiments have been demonstrated by Bisgaard et al. [16] and the PAD was able to provide insights on the time-dependent systems. With these possibilities, one can expect that the well-tested PAD measurement of isotropically distributed molecules plays an important role in the structure determination of molecules.

References

1. A. Stolow, J.G. Underwood, Time-resolved photoelectron spectroscopy of nonadiabatic dynamics in polyatomic molecules. Adv. Chem. Phys. **139**, 497–584 (2008)
2. K.L. Reid, Photoelectron angular distributions. Annu. Rev. Phys. Chem. **54**, 397–424 (2003)
3. D.M. Neumark, Time-resolved photoelectron spectroscopy of molecules and clusters. Annu. Rev. Phys. Chem. **52**, 255–277 (2001)
4. M. Tsubouchi, T. Suzuki, Photoionization of homonuclear diatomic molecules aligned by an intense femtosecond laser pulse. Phys. Rev. A **72**, 022512 (2005)
5. D. Dill, Fixed-molecule photoelectron angular distributions. J. Chem. Phys. **65**, 1130–1133 (1976)
6. F. Rosca-Pruna, M.J.J. Vrakking, Revival structures in picosecond laser-induced alignment of I_2 molecules. I. experimental results. J. Chem. Phys. **116**, 6567–6578 (2002)
7. B. Friedrich, D. Herschbach, Alignment and trapping of molecules in intense laser fields. Phys. Rev. Lett. **74**, 4623–4626 (1995)
8. H. Stapelfeldt, T. Seideman, Colloquium: aligning molecules with strong laser pulses. Rev. Mod. Phys. **75**, 543–557 (2003)
9. I. Thomann, R. Lock, V. Sharma, E. Gagnon, S.T. Pratt, H.C. Kapteyn, M.M. Murnane, W. Li, Direct measurement of the angular dependence of the single-photon ionization of aligned N_2 and CO_2. J. Phys. Chem. A **112**, 9382–9386 (2008)
10. R.R. Lucchese, G. Raseev, V. McKoy, Studies of differential and total photoionization cross sections of molecular nitrogen. Phys. Rev. A **25**, 2572–2587 (1982)
11. R.R. Lucchese, V. McKoy, Studies of differential and total photoionization cross sections of carbon dioxide. Phys. Rev. A **26**, 1406–1418 (1982)
12. F. Kelkensberg, A. Rouzée, W. Siu, G. Gademann, P. Johnsson, M. Lucchini, R.R. Lucchese, M.J.J. Vrakking, XUV ionization of aligned molecules. Phys. Rev. A **84**, 051404 (2011)
13. A.T. Le, R.R. Lucchese, C.D. Lin, Uncovering multiple orbitals influence in high-harmonic generation from aligned N_2. J. Phys. B **42**, 211001 (2009)
14. M. Meckel, D. Comtois, D. Zeidler, A. Staudte, D. Pavicic, H.C. Bandulet, H. Pépin, J.C. Kieffer, R. Dörner, D.M. Villeneuve, P.B. Corkum, Laser-induced electron tunneling and diffraction. Science **320**, 1478–1482 (2008)
15. V. Kumarappan, L. Holmegaard, C. Martiny, C.B. Madsen, T.K. Kjeldsen, S.S. Viftrup, L.B. Madsen, H. Stapelfeldt, Multiphoton electron angular distributions from laser-aligned CS_2 molecules. Phys. Rev. Lett. **100**, 093006 (2008)
16. C.Z. Bisgaard, O.J. Clarkin, G. Wu, A.M.D. Lee, O. Geßner, C.C. Hayden, A. Stolow, Time-resolved molecular frame dynamics of fixed-in-space CS_2 molecules. Science **323**, 1464–1468 (2009)
17. O. Geßner, A.M.D. Lee, J.P. Shaffer, H. Reisler, S.V. Levchenko, A.I. Krylov, J.G. Underwood, H. Shi, A.L.L. East, D.M. Wardlaw, E.T.H. Chrysostom, C.C. Hayden, A. Stolow, Femtosecond multidimensional imaging of a molecular dissociation. Science **311**, 219–222 (2005)
18. C. Jin, A.T. Le, S.-F. Zhao, R.R. Lucchese, C.D. Lin, Theoretical study of photoelectron angular distributions in single-photon ionization of aligned N_2 and CO_2. Phys. Rev. A **81**, 033421 (2010)
19. X.M. Tong, Z.X. Zhao, C.D. Lin, Theory of molecular tunneling ionization. Phys. Rev. A **66**, 033402 (2002)

20. S.-F. Zhao, C. Jin, A.T. Le, T.F. Jiang, C.D. Lin, Determination of structure parameters in strong-field tunneling ionization theory of molecules. Phys. Rev. A **81**, 033423 (2010)
21. D.W. Turner, C. Baker, A.D. Baker, C.R. Brundle, *Molecular Photoelectron Spectroscopy: A Handbook of He 584 Å Spectra* (Wiley, London, 1970)
22. B.K. McFarland, J.P. Farrell, P.H. Bucksbaum, M. Gühr, High harmonic generation from multiple orbitals in N_2. Science **322**, 1232–1235 (2008)
23. W. Boutu, S. Haessler, H. Merdji, P. Breger, G. Waters, M. Stankiewicz, L.J. Frasinski, R. Taieb, J. Caillat, A. Maquet, P. Monchicourt, B. Carre, P. Saliéres, Coherent control of attosecond emission from aligned molecules. Nature Phys. **4**, 545–549 (2008)
24. A.T. Le, R.R. Lucchese, M. T. lee, and C. D. Lin. Probing molecular frame photoionization via laser generated high-order harmonics from aligned molecules. Phys. Rev. Lett. **102**, 203001 (2009)
25. X. Zhou, R. Lock, W. Li, N. Wagner, M.M. Murnane, H.C. Kapteyn, Molecular recollision interferometry in high harmonic generation. Phys. Rev. Lett. **100**, 073902 (2008)
26. A.T. Le, R.R. Lucchese, S. Tonzani, T. Morishita, C.D. Lin, Quantitative rescattering theory for high-order harmonic generation from molecules. Phys. Rev. A **80**, 013401 (2009)
27. E. Gagnon, P. Ranitovic, X.-M. Tong, C.L. Cocke, M.M. Murnane, H.C. Kapteyn, A.S. Sandhu, Soft X-ray-driven femtosecond molecular dynamics. Science **317**, 1374–1378 (2007)

Chapter 8
Summary

High-order harmonic generation (HHG) is a dramatic nonlinear process when atoms or molecules are exposed to an intense infrared laser pulse. In this process, the fundamental laser light is efficiently converted to an extreme ultraviolet (XUV) or soft X-ray light. Its potential has been shown to probe the electronic structure of targets and the time-resolved molecular structure, ionize aligned molecules, and produce an attosecond pulse train or isolated attosecond pulse. In this thesis, I have established a complete model for the HHG in a gaseous medium by incorporating a quantitative rescattering (QRS) theory [1–3] with the standard macroscopic propagation theory. Here I will summarize the main achievements in this thesis.

1. Numerical modeling of macroscopic HHG[4–6]
To describe a HHG process completely, one needs to consider both the single-atom or single-molecule response and the macroscopic response of medium. The QRS theory is an efficient approach to calculate the single-atom or single-molecule induced dipole, which has the comparable accuracy with solving the time-dependent Schrödinger equation (TDSE). Induced dipoles are then fed into the propagation equation of high-harmonic field. Meanwhile, plasma, dispersion and Kerr effects are included for the propagation of fundamental laser field in the medium. The further free propagation of high harmonics emitted from the exit plane of a gas jet can be handled by the Hankel transformation. Two types of spatial beams for the fundamental laser, Gaussian and truncated Bessel beams, are incorporated into the propagation model, which are mainly dealing with multi-cycle and few-cycle laser pulses, respectively. The model has been checked against the TDSE results by comparing both the magnitude and the phase of macroscopic HHG.

2. Quantitative comparison with measured HHG [7–10]
Since the HHG is a coherent process, there are many factors which can influence the HHG spectrum. The direct comparison with the measured HHG spectrum for atoms is rare, and it has not been done for molecules as far as my knowledge goes. In usual studies of the single-atom or single-molecule HHG, laser parameters such as

C. Jin, *Theory of Nonlinear Propagation of High Harmonics Generated in a Gaseous Medium*, Springer Theses, DOI: 10.1007/978-3-319-01625-2_8,

intensity, pulse duration, wavelength and carrier-envelope phase (CEP), are mainly concerned. Besides these parameters, I also include position and width of a gas jet, gas pressure, spatial beam modes and detecting system in the model. The more experimental parameters can be specified, the better simulations can be resulted from the model. For a few examples with the well specified experimental conditions, I am able to show the good agreement between the measured HHG spectrum and the simulation over a broad photon-energy region for both isotropic and aligned N_2 and CO_2 molecules. According to my knowledge, this is the first time that the measured HHG spectrum of molecules can be described quantitatively by the theory. I also show the good quantitative agreement between measured HHG spectra of Ar and simulations for different wavelengths. For the HHG spectrum of Xe measured using a mid-infrared laser pulse, the model is able to predict the multi-electron effect and the continuum structure.

3. Factorization of macroscopic HHG [4, 5, 7]
In the QRS theory, HHG from a single atom or a single molecule can be separated as an electron returning wave packet and a photorecombination (PR) transition dipole. In this thesis, I have proven numerically that the macroscopic HHG can be written as a product of a "macroscopic wave packet" (MWP) and a PR transition dipole. For a molecular target, the PR transition dipole is replaced by an alignment-averaged one. The PR transition dipole reflects the electronic structure of a target, i.e., the property of a target only. While laser and medium propagation effects can all be included in the MWP. For the same target, all variations of HHG spectra by changing the experimental conditions can be attributed to the differences in MWP. This factorization (or the separable approximation) of HHG after the macroscopic propagation provides a necessary theoretical basis for extracting the electronic structure of molecular orbital from the measured HHG spectrum.

4. Spatial filtering in the far field for isolated attosecond pulse generation [10, 11]
The continuum harmonics in the spectrum (in frequency domain) are generally used to produce an isolated attosecond pulse (IAP) (in time domain). However, it is challenged to measure the duration of an IAP experimentally. To make a connection between an IAP and the continuum harmonics, alternatively it has to rely on the theoretical model. In this thesis, I have investigated the continuum structure in the HHG spectrum of Xe, measured with a CEP-not-stabilized few-cycle mid-infrared laser pulse. By using the time-frequency (or wavelet) analysis, I am able to show that the reshaping of fundamental laser field is responsible for the continuous harmonics. And then I have suggested an approach to create an IAP by using a filter centered on axis to select the high harmonics in the far field with different divergence. This approach has been tested for different CEPs.

5. Multiple orbital contribution in the HHG of aligned molecules [8, 9]
The first step of the HHG process is tunneling ionization. Since the tunneling ionization rate depends on the ionization potential exponentially, electrons are usually ion-

ized from the outmost molecular orbital (or the highest-occupied molecular-orbital, HOMO) only. However, the ionization rate also depends on the symmetry of molecular orbital. At some alignment angles, ionization rate of the outmost molecular orbital becomes small, while it becomes large for some inner molecular orbitals. In this case, the HHG process could occur in both outmost and inner molecular orbitals. There are two mostly studied molecules, N_2 and CO_2. HOMO-1 contributes to the HHG when the laser polarization is perpendicular to the molecular axis of N_2. For CO_2, the contribution of HOMO-2 is presented in the HHG when the laser polarization is parallel to the molecular axis. In this thesis, I have shown that the HOMO-1 contribution of N_2 in the HHG spectrum over a broad region of the photon energy can be controlled by laser intensity. And the minimum in the HHG spectrum of CO_2 can be easily influenced by many factors due to the interference between HOMO and HOMO-2. This can explain why the minima measured in different laboratories may vary.

6. Photoionization of aligned molecules with HHG light [12]
The traditional photoionization is a basic tool to investigate the molecular structure. In the past, this was mostly done for isotropically distributed molecules by using the synchrotron radiation. Infrared femtosecond lasers, which are widely available in many labs change this situation in two aspects. They can be used to transiently align molecules, and also be used to produce the high harmonics, which can serve as XUV or soft X-ray light. Using infrared laser facilities only, one can photoionize aligned molecules with the HHG light. The rich structure of photoelectron angular distribution (PAD) for fixed-in-space molecules can then be observed. In this chapter, I have presented the PADs for aligned N_2 and CO_2 molecules in the laboratory frame by using the well-established photoionization theory [13, 14]. These calculated results are expected to compare with future measurements directly.

In brief, I have established an *ab initio* model to describe the HHG completely. A few interesting issues in the field have been touched in this thesis. All these studies strengthen the theoretical basis for the applications of HHG, and they are also helpful for experimentalists in the future.

References

1. T. Morishita, A.T. Le, Z. Chen, C.D. Lin, Accurate retrieval of structural information from laser-induced photoelectron and high-order harmonic spectra by few-cycle laser pulses. Phys. Rev. Lett. **100**, 013903 (2008)
2. A.T. Le, R.R. Lucchese, S. Tonzani, T. Morishita, C.D. Lin, Quantitative rescattering theory for high-order harmonic generation from molecules. Phys. Rev. A **80**, 013401 (2009)
3. C.D. Lin, A.T. Le, Z. Chen, T. Morishita, R. Lucchese, Strong-field rescattering physics–self-imaging of a molecule by its own electrons. J. Phys. B **43**, 122001 (2010)
4. C. Jin, A.T. Le, C.D. Lin, Retrieval of target photorecombination cross sections from high-order harmonics generated in a macroscopic medium. Phys. Rev. A **79**, 053413 (2009)
5. C. Jin, A.T. Le, C.D. Lin, Medium propagation effects in high-order harmonic generation of Ar and N_2. Phys. Rev. A **83**, 023411 (2011)

6. C. Jin, C.D. Lin, Comparison of high-order harmonic generation of Ar using truncated bessel and gaussian beams. Phys. Rev. A **85**, 033423 (2012)
7. C. Jin, H.J. Wörner, V. Tosa, A.T. Le, J.B. Bertrand, R.R. Lucchese, P.B. Corkum, D.M. Villeneuve, C.D. Lin, Separation of target structure and medium propagation effects in high-harmonic generation. J. Phys. B **44**, 095601 (2011)
8. C. Jin, A.T. Le, C.D. Lin, Analysis of effects of macroscopic propagation and multiple molecular orbitals on the minimum in high-order harmonic generation of aligned CO_2. Phys. Rev. A **83**, 053409 (2011)
9. C. Jin, J.B. Bertrand, R.R. Lucchese, H.J. Wörner, P.B. Corkum, D.M. Villeneuve, A.T. Le, C.D. Lin, Intensity dependence of multiple-orbital contributions and shape resonance in high-order harmonic generation of aligned N_2 molecules. Phys. Rev. A **85**, 013405 (2012)
10. C. Trallero-Herrero, C. Jin, B.E. Schmidt, A.D. Shiner, J.-C. Kieffer, P.B. Corkum, D.M. Villeneuve, C.D. Lin, F. Légaré, A.T. Le, Generation of broad XUV continuous high harmonic spectra and isolated attosecond pulses with intense mid-infrared lasers. J. Phys. B **45**, 011001 (2012)
11. C. Jin, A.T. Le, C.A. Trallero-Herrero, C.D. Lin, Generation of isolated attosecond pulses in the far field by spatial filtering with an intense few-cycle mid-infrared laser. Phys. Rev. A **84**, 043411 (2011)
12. C. Jin, A.T. Le, S.-F. Zhao, R.R. Lucchese, C.D. Lin, Theoretical study of photoelectron angular distributions in single-photon ionization of aligned N_2 and CO_2. Phys. Rev. A **81**, 033421 (2010)
13. R.R. Lucchese, G. Raseev, V. McKoy, Studies of differential and total photoionization cross sections of molecular nitrogen. Phys. Rev. A **25**, 2572–2587 (1982)
14. R.R. Lucchese, V. McKoy, Studies of differential and total photoionization cross sections of carbon dioxide. Phys. Rev. A **26**, 1406–1418 (1982)

Appendix A
Abbreviations

HHG	High-order harmonic generation
QRS	Quantitative rescattering
TDSE	Time-dependent schrödinger equation
SFA	Strong-field approximation
MWP	Macroscopic wave packet
SAE	Single-active electron
CM	Cooper minimum
ADK	Ammosov–Delone–Krainov
IR	Infrared
MIR	Mid-infrared
NIR	Near-infrared
XUV	Extreme ultraviolet
UV	Ultraviolet
APT	Attosecond pulse train
IAP	Isolated attosecond pulse
TFR	Time-frequency representation
MO-ADK	Molecular Ammosov–Delone–Krainov
HOMO	Highest-occupied molecular-orbital
PRCS	Photorecombination cross section
PR	Photorecombination
PICS	Photoionization cross section
PI	Photoionization
PAD	Photoelectron angular distribution
MF-PAD	Molecular frame photoelectron angular distribution
LF-PAD	Photoelectron angular distribution in the laboratory frame
DCS	Differential cross section
IDAD	Integrated detector angular distribution
ITAD	Integrated target angular distribution
CASSCF	Complete-active-space self-consistent field
CI	Configuration interaction
CEP	Carrier-envelope phase
FWHM	Full width at half maximum
3-D	Three-dimensional
TB	Truncated bessel
TB-1	Type-1 bessel
TB-2	Type-2 bessel

C. Jin, *Theory of Nonlinear Propagation of High Harmonics Generated in a Gaseous Medium*, Springer Theses, DOI: 10.1007/978-3-319-01625-2,

HATI	High-energy above-threshold ionization
NSDI	Nonsequential double ionization
GDD	Group delay dispersion
RRPA	Relativistic random-phase approximation
RABITT	Reconstruction of attosecond beating by interference of two-photon transitions
COLTRIMS	Cold target recoil ion momentum spectroscopy
HCF	Hollow-core fiber

Appendix B
Theory of Alignment for Linear Molecules

When linear molecules are placed in a femtosecond laser pulse (this is usually called as the pump laser), a rotational wave packet can be excited. At later times the molecules would be aligned or antialigned when the wave packet undergoes "rotational revival" [1, 2]. To calculate the degree of alignment, or the molecular angular distribution, each molecule can be treated as a rigid rotor [3, 4]. The evolution of rotational wave packet with initial state $\Psi_{JM}(\theta, \varphi, t = -\infty)$ in a linearly polarized laser field can be described by the time-dependent Schrödinger equation (atomic units), which is given by

$$i\frac{\partial \Psi_{JM}(\theta, \varphi, t)}{\partial t} = \left[B\mathbf{J}^2 - \frac{E(t)^2}{2}(\alpha_{\parallel} \cos^2 \theta + \alpha_{\perp} \sin^2 \theta) \right] \Psi_{JM}(\theta, \varphi, t), \quad \text{(B.1)}$$

where $\mathbf{J}$ is the angular momentum operator, B is the rotational constant, $\alpha_{\parallel}$ and $\alpha_{\perp}$ are anisotropic polarizabilities in parallel and perpendicular directions with respect to the molecular axis, respectively. These molecular properties for N_2, O_2 and CO_2 molecules are shown in Table B.1. The electric field of pump laser, $E(t)$, in Eq. (B.1), has a Gaussian envelope:

$$E(t) = E_0 e^{-(2\ln 2)t^2/\tau_w^2} \cos(\omega_0 t), \quad \text{(B.2)}$$

where E_0 is the peak field, τ_w and ω_0 are the pulse duration (full width at half maximum, FWHM) and the frequency of pump laser, respectively. Note that Eq. (B.1) is expressed in the molecular (or body-fixed) frame.

For each initial rotational state $|JM\rangle$ (up to $J = 40$), Eq. (B.1) could be solved independently using the split-operator method [7, 8]. After the pump laser is turned off, the rotational wave packet will continue to propagate in free space,

$$\Psi_{JM}(t) = \sum_{J'} a_{J'} e^{-iE_{J'}t} |J'M\rangle, \quad \text{(B.3)}$$

C. Jin, *Theory of Nonlinear Propagation of High Harmonics Generated in a Gaseous Medium*, Springer Theses, DOI: 10.1007/978-3-319-01625-2,

Table B.1 Molecular properties for N_2, O_2 and CO_2

Molecule	B (cm^{-1})	$\alpha_{\parallel}$ (Å^3)	$\alpha_{\perp}$ (Å^3)
N_2	1.989	2.38	1.45
O_2	1.438	2.35	1.21
CO_2	0.39	4.05	1.95

B is the rotational constant, $\alpha_{\parallel}$ and $\alpha_{\perp}$ are parallel and perpendicular polarizabilities, respectively. The data are taken from [5, 6]

where $E_{J'}$ are energy eigenvalues, $|J'M\rangle$ are spherical harmonics, and coefficients of $a_{J'}$ are determined at the moment when the pump laser is turned off.

Assuming a Boltzman distribution of rotational levels at the initial time, the time-dependent alignment at a given temperature can be obtained by

$$\rho(\theta, t) = \sum_{JM} \omega_{JM} |\Psi_{JM}(\theta, \varphi, t)|^2, \tag{B.4}$$

where ω_{JM} is the weight according to the Boltzman distribution. To determine ω_{JM}, the nuclear statistics and the symmetry of total electronic wave function must be taken into account properly. Note that for linear molecules the angular distribution (or the alignment) does not depend on the azimuthal angle φ in the frame attached to the pump laser field, and it only depends on the angle θ between the molecular axis and the polarization direction of pump laser.

References

1. A.D. Bandrauk, J. Ruel, Charge-resonance-enhanced ionization of molecular ions in intense laser pulses: Geometric and orientation effects. Phys. Rev. A **59**, 2153–2162 (1999)
2. T. Seideman, Revival structure of aligned rotational wave packets. Phys. Rev. Lett. **83**, 4971–4974 (1999)
3. H. Stapelfeldt, T. Seideman, Colloquium: aligning molecules with strong laser pulses. Rev. Mod. Phys. **75**, 543–557 (2003)
4. J. Ortigoso, M. Rodríguez, M. Gupta, B. Friedrich, Time evolution of pendular states created by the interaction of molecular polarizability with a pulsed nonresonant laser field. J. Chem. Phys. **110**, 3870–3875 (1999)
5. J.O. Hirschfelder, C.F. Curtiss, R.B. Bird, *Molecular Theory of Gases and Liquids* (Wiley, New York, 1954)
6. P. J. Linstrom, W.G. Mallard, in *NIST Chemistry WebBook, NIST Standard Reference Database* vol. 69, ed. by P. J. Linstrom, W.G. Mallard (National Institute of Standards and Technology, Gaithersburg, 2006), p. 20899
7. X.M. Tong, S.I. Chu, Theoretical study of multiple high-order harmonic generation by intense ultrashort pulsed laser fields: a new generalized pseudospectral time-dependent method. Chem. Phys. **217**, 119–130 (1997)
8. X.M. Tong, S.I. Chu, Time-dependent approach to high-resolution spectroscopy and quantum dynamics of Rydberg atoms in crossed magnetic and electric fields. Phys. Rev. A **61**, 031401 (2000)

Appendix C
Photorecombination Transition Dipole

C.1 Photorecombination Transition Dipole of Atomic Targets

Photoionization transition dipole from an initial bound state Ψ_i to the final continuum state $\Psi_{\vec{k}}^{(-)}$ due to a linearly polarized light is [1]

$$d_{\vec{k},\vec{n}}(\omega) = \langle \Psi_i | \vec{r} \cdot \vec{n} | \Psi_{\vec{k}}^{(-)} \rangle. \tag{C.1}$$

Here $\vec{n}$ is the direction of light polarization and $\vec{k}$ is the momentum of ejected photoelectron. The photoionization differential cross section (DCS) is proportional to the modulus square of this transition dipole (in the length form):

$$\frac{d^2\sigma^I}{d\Omega_{\vec{k}} d\Omega_{\vec{n}}} = \frac{4\pi^2 \omega k}{c} |\langle \Psi_i | \vec{r} \cdot \vec{n} | \Psi_{\vec{k}}^{(-)} \rangle|^2, \tag{C.2}$$

where $k^2/2 + I_p = \omega$ (atomic units) with I_p being the ionization potential, ω the photon energy, and c the speed of light. The continuum wave function $\Psi_{\vec{k}}^{(-)}(\vec{r})$ satisfies the stationary Schrödinger equation

$$\left[-\frac{\nabla^2}{2} + V(r) - \frac{k^2}{2} \right] \Psi_{\vec{k}}^{(-)}(\vec{r}) = 0 \tag{C.3}$$

where the spherically symmetric model potential $V(r)$ in Eqs. (2.34) and (2.35) is also used.

The more relevant process to the HHG is its time-reversed one-photon photorecombination process. The photorecombination DCS can be written as

$$\frac{d^2\sigma^R}{d\Omega_{\vec{n}} d\Omega_{\vec{k}}} = \frac{4\pi^2 \omega^3}{c^3 k} |\langle \Psi_i | \vec{r} \cdot \vec{n} | \Psi_{\vec{k}}^{(+)} \rangle|^2. \tag{C.4}$$

C. Jin, *Theory of Nonlinear Propagation of High Harmonics Generated in a Gaseous Medium*, Springer Theses, DOI: 10.1007/978-3-319-01625-2,

In comparison with the photoionization DCS in Eq. (C.2), the continuum state is taken as the outgoing scattering wave $\Psi_{\vec{k}}^{(+)}$ instead of an incoming wave $\Psi_{\vec{k}}^{(-)}$, and there is also a different overall factor. In fact, the photoionization and photorecombination DCS's are related by

$$\frac{d^2\sigma^R}{\omega^2 d\Omega_{\hat{n}} d\Omega_{\vec{k}}} = \frac{d^2\sigma^I}{c^2k^2 d\Omega_{\hat{k}} d\Omega_{\vec{n}}}, \tag{C.5}$$

which follows the principle of the detailed balancing for direct and time-reversed processes [2].

C.2 Doubly Differential PICS in the Molecular Frame

Similar to Eq. (C.2), the doubly differential photoionization cross section (PICS) in the molecular (or body-fixed) frame is [1, 3–5]

$$\frac{d^2\sigma^I}{d\Omega_{\hat{k}} d\Omega_{\hat{n}}} = \frac{4\pi^2\omega}{c}|I_{\vec{k},\hat{n}}|^2. \tag{C.6}$$

Here the dipole matrix elements from an initial bound state Ψ_i to the continuum state $\Psi_{f,\vec{k}}^{(-)}$ due to the linearly polarized light in the dipole length approximation are

$$I_{\vec{k},\hat{n}} = (k)^{1/2}\langle\Psi_i|\vec{r}\cdot\hat{n}|\Psi_{f,\vec{k}}^{(-)}\rangle, \tag{C.7}$$

where $\hat{n}$ is the polarization direction of light, and $\vec{k}$ the momentum of photoelectron.

To treat the angular dependence of PICS on the target orientation, the dipole matrix elements are expanded in terms of spherical harmonics

$$I_{\vec{k},\hat{n}} = (\frac{4\pi}{3})^{1/2}\sum_{lm\mu} I_{lm\mu} Y_{lm}^*(\Omega_{\hat{k}}) Y_{l\mu}^*(\Omega_{\hat{n}}). \tag{C.8}$$

Partial-wave matrix elements are given by

$$I_{lm\mu} = (k)^{1/2}\langle\Psi_i|r_\mu|\Psi_{f,klm}^{(-)}\rangle, \tag{C.9}$$

where

$$r_\mu = \begin{cases} \mp(x \pm iy)/2^{1/2} & \mu = \pm 1, \\ z & \mu = 0. \end{cases} \tag{C.10}$$

In the calculation, the initial bound state Ψ_i is obtained from the MOLPRO code [6] within the valence complete-active-space self-consistent field (CASSCF) method. Based on the frozen-core approximation the final state $\Psi_{f,\vec{k}}^{(-)}$ is then described in a single-channel approximation where the wave function of ionic core is given by a valence complete active space configuration interaction (CI) wave function obtained using the same bound orbitals as in the initial state. It has the form

$$\Psi^{(-)}_{f,\vec{k}} = \mathbf{A}[\Phi\phi^{(-)}_{\vec{k}}(\vec{r})], \tag{C.11}$$

where Φ is the correlated $N-1$ electron ionic-core wave function, $\phi^{(-)}_{\vec{k}}(\vec{r})$ is the one-electron continuum wave function, and operator $\mathbf{A}$ performs the appropriate antisymmetrization of spin and spatial symmetry adaptation of the product of the ionic core and continuum wave functions. Note that it is possible to use ionic orbitals, however, CI calculations are performed in both initial and final states, the choice of orbitals does not affect much the final results. For the valence ionization, the position of one-electron continuum resonances has been reproduced quite well using initial state orbitals. So starting with initial state orbitals and using CI wave functions give quite reliable results.

The Schrödinger equation for the remaining continuum electron is then (in atomic units)

$$\left[-\frac{1}{2}\nabla^2 - \frac{1}{r} + V(\vec{r}) - \frac{k^2}{2}\right]\phi^{(-)}_{\vec{k}}(\vec{r}) = 0, \tag{C.12}$$

where $V(\vec{r})$ is the short-range portion of electron-molecular-ion interaction. Note that the potential is not spherically symmetric for molecules. Equation (C.12) is then solved by using the iterative Schwinger variational method. The continuum wave function is expanded in terms of partial waves as

$$\phi^{(-)}_{\vec{k}}(\vec{r}) = (\frac{2}{\pi})^{1/2}\sum_{l=0}^{l_p}\sum_{m=-l}^{l} i^l \phi^{(-)}_{klm}(\vec{r})Y^*_{lm}(\Omega_{\hat{k}}), \tag{C.13}$$

where an infinite sum over l has been truncated at $l = l_p$. In the calculation, $l_p = 11$. Once $\phi^{(-)}_{klm}(\vec{r})$ is obtained, $\Psi^{(-)}_{f,klm}$ in Eq. (C.9) can be calculated straightforwardly through Eq. (C.11). Note that the continuum wave function is constructed to be orthogonal to strongly occupied orbitals. This avoids spurious singularities which can occur when the scattering from correlated targets is considered.

Here the method used to compute the scattering potential $V(\vec{r})$ in Eq. (C.12) is described. First, the electronic Hamiltonian is written as

$$H = \sum_{i=1}^{N} h(i) + \sum_{i<j}^{N}\frac{1}{r_{ij}}, \tag{C.14}$$

with

$$h(i) = -\frac{\nabla_i^2}{2} - \sum_a \frac{Z_a}{r_{ia}}, \tag{C.15}$$

where Z_a are the nuclear charges, and N is the number of electrons. Then the single-particle equation for the continuum electron is obtained from

$$\langle \delta\Psi^{(-)}_{f,\vec{k}} | H - E | \Psi^{(-)}_{f,\vec{k}} \rangle = 0, \tag{C.16}$$

where $\delta\Psi^{(-)}_{f,\vec{k}}$ is written as in Eq. (C.11), with $\phi^{(-)}_{\vec{k}}(\vec{r})$ replaced by $\delta\phi^{(-)}_{\vec{k}}(\vec{r})$. By requiring this equation to be satisfied for all possible $\delta\Psi^{(-)}_{f,\vec{k}}$ [or $\delta\phi^{(-)}_{\vec{k}}(\vec{r})$], one obtains a nonlocal optical potential that can be written in the form of a Phillips-Kleinman pseudopotential.

A single-center expansion approach is used to evaluate all required matrix elements. In other words, all functions, including the scattering wave function, occupied orbital and potential are expanded about a common origin, which is the center of mass of the molecule, as a sum of spherical harmonics times radial functions

$$F(\vec{r}) = \sum_{l=0}^{l_{\max}} \sum_{m=-l}^{l} f_{lm}(r) Y_{lm}(\theta, \phi). \tag{C.17}$$

With this expansion, the angular integration can be done analytically and all three-dimensional integrals reduce to a sum of radial integrals, which are computed on a radial grid. Typically, $l_{\max} = 60$–85.

If one is dealing with the electron ionization from inner molecular orbitals, i.e., not the highest-occupied molecular-orbital (HOMO), but rather the HOMO-1 and the HOMO-2, it can still be done in the same manner, except that the ionic-core state Φ employed in Eq. (C.11) needs to be replaced by the excited ion state, which corresponds to the electron ionization from the HOMO-1 or the HOMO-2 orbital. Furthermore, the present single-channel formalism can be extended to the coupled-multichannel calculations to account for additional electron correlation effects. The calculations in this thesis are limited to the single-channel approximation.

As mentioned in the quantitative rescattering (QRS) theory, see Sect. 2.2.3, photorecombination transition dipoles are involved in the process of high-order harmonic generation. It only has a difference in the sign of phase, but has the same magnitude as compared to the photoionization transition dipole [1]. To perform the QRS calculation, the returning electron wave packet is usually obtained by using the strong-field approximation for simplicity, where the photorecombination transition dipole is either a pure real or pure imaginary number. And then the exact photorecombination transition dipole is incorporated with the wave packet to obtain the induced dipole moment.

C.3 Alignment Dependence of Integrated PICS

Doubly differential PICS in the molecular frame is given in Eq. (C.6), but for a given application one may need averaged PICSs as suggested by Wallace and Dill [7]. One such averaged distribution is the integrated detector angular distribution (IDAD), which corresponds to the experiments where the target orientation is fixed in space and the PICS is integrated over all possible emission directions of the photoelectron. For linear molecules, the integrated cross section depends only on the alignment angle θ due to the symmetry of molecules and the IDAD can be expressed in the molecular frame as

$$\sigma(\theta, \omega) = \frac{d\sigma}{d\Omega_{\hat{n}}} = \int \frac{4\pi^2\omega}{c} |I_{\vec{k},\hat{n}}|^2 d\Omega_{\hat{k}}. \tag{C.18}$$

Equation (C.18) can also be found in the form of [3, 7]

$$\frac{d\sigma}{d\Omega_{\hat{n}}} = \frac{\sigma_{\text{tot}}}{4\pi}[1 + \beta_{\hat{n}} P_2(\cos\theta)], \tag{C.19}$$

where σ_{tot} is the total PICS averaged over all alignments and photoelectron directions, $P_2(\cos\theta)$ is the Legendre polynomial of degree 2 and $\beta_{\hat{n}}$ is the asymmetry parameter.

In the experiments, pump laser is used to create transiently aligned molecular sample, and then XUV or soft X-rays probe ionizes the molecules. The time-dependent alignment distribution is obtained by solving the time-dependent Schrödinger equation. If the polarizations of pump and probe lasers are parallel, the detected experimental signal in terms of pump-probe time delay τ can be written as

$$Y(\omega, \tau) \propto \int_0^{\pi} \sigma(\theta, \omega)\rho(\theta, \tau) \sin\theta d\theta. \tag{C.20}$$

Without the pump-probe scheme, the molecules are distributed randomly, and the angular distribution of $\rho(\theta, \tau)$ in Equation (C.20) is a constant. Equation (C.20) actually gives one the total cross section σ_{tot}.

C.4 PAD in the Laboratory Frame

The PICS in the molecular frame is given in Eq. (C.6), and the doubly differential PICS in the laboratory frame can be expressed as

$$\frac{d^2\sigma}{d\Omega_{\hat{k}'} d\Omega_{\hat{n}'}} = \frac{4\pi^2\omega}{c} |I_{\vec{k}',\hat{n}'}|^2, \tag{C.21}$$

where $\hat{n}'$ and $\vec{k}'$ are the polarization direction of pump laser and the momentum of photoelectron in the laboratory frame, respectively. Assuming that the molecular axis is aligned at an arbitrary angle $\hat{\mathbf{R}} \equiv (\theta, \varphi)$ with respect to the polarization direction of pump laser. In other words, $\hat{\mathbf{R}}$ is the Euler angle of molecular frame with respect to the laboratory frame. Dipole matrix elements in Eq. (C.8) can be rewritten in the laboratory frame as

$$I_{\vec{k}',\hat{n}'} = (\frac{4\pi}{3})^{1/2} \sum_{lm\mu} I_{lm\mu} \sum_{m'=-l}^{l} D^{l}_{mm'}(\hat{\mathbf{R}}) Y^{*}_{lm'}(\theta_{k'}, \varphi_{k'}) \sum_{\mu'=-l}^{l} D^{l}_{\mu\mu'}(\hat{\mathbf{R}}) Y^{*}_{l\mu'}(\theta_{n'}, \varphi_{n'}), \tag{C.22}$$

with $D^{l}_{mm'}(\hat{\mathbf{R}})$ and $D^{l}_{\mu\mu'}(\hat{\mathbf{R}})$ being the rotation matrices. In Eq. (C.22), $\theta_{k'}$ and $\varphi_{k'}$ are polar and azimuthal angles of the photoelectron in the laboratory frame, respectively, $\theta_{n'} = 0°$ and $\varphi_{n'} = 0°$ in the laboratory frame. The PICS of Eq. (C.21) in the laboratory frame is an explicit function of $\theta_{k'}$ and $\varphi_{k'}$ for the alignment angle $\hat{\mathbf{R}}$.

In the laboratory frame, taking into account the molecular distribution with respect to the polarization direction of pump laser described by the angle θ, the PICS in Eq. (C.21) must be integrated over the azimuthal angle φ. Finally, one can obtain the PICS for all molecules with a fixed alignment angle θ, which depends on the photoelectron emission angle $\theta_{k'}$,

$$\sigma'(\theta, \omega, \theta_{k'}) = \int_0^{2\pi} \frac{d^2\sigma}{d\Omega_{\hat{k}'} d\Omega_{\hat{n}'}} (\theta, \theta_{k'}, \varphi_{k'} - \varphi) d\varphi. \tag{C.23}$$

The actual experimentally observed photoelectron angular distribution in the laboratory frame (LF-PAD) corresponds to the average of PICS in Eq. (C.23) accounting for the molecular distribution in space. The time dependent angular distribution of $\rho(\theta, \tau)$ can be calculated by Eq. (B.4), and the analytical form of LF-PAD, which can be compared with the experimental photoelectron spectra directly, is expressed as

$$Y'(\omega, \theta_{k'}, \tau) \propto \int_0^{\pi} \sigma'(\theta, \omega, \theta_{k'}) \rho(\theta, \tau) \sin\theta d\theta. \tag{C.24}$$

The polarizations of pump and probe lasers are parallel in Eq. (C.24).

As suggested by Wallace and Dill [7], another averaged PICS is the integrated target angular distribution (ITAD), which corresponds to PI experiments where the target orientation is not resolved. For isotropically distributed molecules, the angular distribution of $\rho(\theta, \tau)$ is a constant, and Eq. (C.24) has the form [3, 7]

$$\frac{d\sigma}{d\Omega_{\hat{k}'}} = \frac{\sigma_{\mathrm{tot}}}{4\pi}[1 + \beta_{\hat{k}'} P_2(\cos\theta_{k'})], \tag{C.25}$$

where $\beta_{\hat{k}'}$ is the photoelectron asymmetry parameter.

References

1. A.T. Le, R.R. Lucchese, S. Tonzani, T. Morishita, C.D. Lin, Quantitative rescattering theory for high-order harmonic generation from molecules. Phys. Rev. A **80**, 013401 (2009)
2. L.D. Landau, E.M. Lifshitz, *Quantum Mechanics: Nonrelativistic Theory* (Pergamon Press, New York, 1965)
3. R.R. Lucchese, G. Raseev, V. McKoy, Studies of differential and total photoionization cross sections of molecular nitrogen. Phys. Rev. A **25**, 2572–2587 (1982)
4. R.R. Lucchese, V. McKoy, Studies of differential and total photoionization cross sections of carbon dioxide. Phys. Rev. A **26**, 1406–1418 (1982)
5. C. Jin, A.T. Le, S.-F. Zhao, R.R. Lucchese, C.D. Lin, Theoretical study of photoelectron angular distributions in single-photon ionization of aligned N_2 and CO_2. Phys. Rev. A **81**, 033421 (2010)
6. H.-J. Werner et al., *MOLPRO, Version 2002.6, A Package of Ab Initio Programs* (Birmingham, 2003)
7. S. Wallace, D. Dill, Detector integrated angular distribution: Chemisorption-site geometry, axial-recoil photofragmentation, and molecular-beam orientation. Phys. Rev. B **17**, 1692–1699 (1978)

Appendix D
Spatial Mode of Laser Beam: Gaussian Beam Versus Truncated Bessel Beam

D.1 Gaussian Beam

In optics, a Gaussian beam is a beam of electromagnetic radiation whose transverse electric field is well approximated by the Gaussian function. For a Gaussian beam, the complex electric field is given by

$$E_{\text{gau}}(r,z) = \frac{bE_0}{b+2iz}\exp\left(-\frac{kr^2}{b+2iz}\right) = |E_{\text{gau}}(r,z)|e^{i\phi_{laser}(r,z)}. \tag{D.1}$$

Here E_0 is the laser peak field at the focus, ω_0 is the central frequency, $k = \omega_0/c = 2\pi/\lambda_0$ is the wave vector. As shown in Fig. D.1, the geometry and the behavior of a Gaussian beam are governed by a set of beam parameters. The spot size $w(z)$ is at a minimum value w_0 along z-axis, which is the beam waist. At a distance z, the variation of spot size is given by

$$w(z) = w_0\sqrt{1+\left(\frac{z}{z_R}\right)^2}, \tag{D.2}$$

where

$$z_R = \frac{\pi w_0^2}{\lambda_0}, \tag{D.3}$$

is called the Rayleigh range. b is the confocal parameter (the depth of focus) given by twice the distance along z-axis for the beam to expand from its minimum cross sectional area at $z = 0$ to twice this area, i.e., $b = 2z_R$.

From Eq. (D.1), one can also express the intensity and the phase of a Gaussian beam explicitly

$$I(r,z) = |E_{\text{gau}}(r,z)|^2 = \frac{I_0}{1+(2z/b)^2}\cdot\exp\left[-\left(\frac{r}{w_0}\right)^2\cdot\frac{2}{1+(2z/b)^2}\right], \tag{D.4}$$

C. Jin, *Theory of Nonlinear Propagation of High Harmonics Generated in a Gaseous Medium*, Springer Theses, DOI: 10.1007/978-3-319-01625-2,

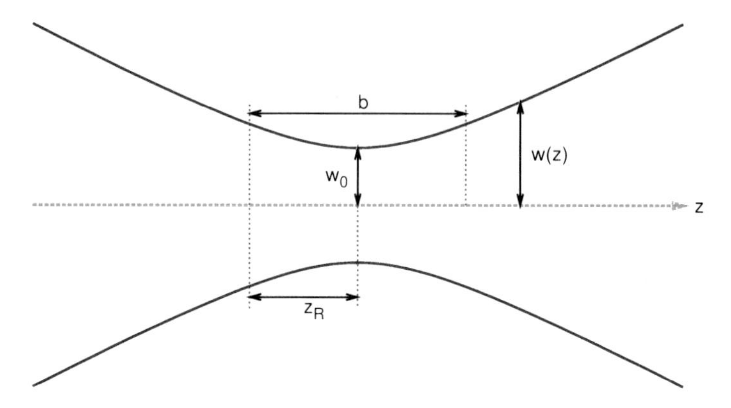

Fig. D.1 Schematic diagram of *a* Gaussian beam. Beam width $w(z)$ as a function of the axial distance z; w_0 beam waist; b confocal parameter, twice of Rayleigh range z_R. Reproduced from [1]

$$\begin{aligned}\phi_{\text{laser}}(r,z) &= -\tan^{-1}(\frac{2z}{b}) + \frac{2kr^2z}{b^2+4z^2}\\ &= -\tan^{-1}(\frac{2z}{b}) + (\frac{r}{w_0})^2 \cdot \frac{(2z/b)}{1+(2z/b)^2}.\end{aligned} \tag{D.5}$$

The geometric phase due to laser focusing is given by ϕ_{laser}. $\tan^{-1}(2z/b)$ is the *Gouy phase*, which results in a phase shift of π relative to a plane wave as the laser passes through the focus. As shown in Eqs. (D.4) and (D.5), if the propagation distance z is scaled by the confocal parameter b, and the radial distance r is scaled by the beam waist w_0, the intensity and the phase stay the same.

D.2 Truncated Bessel Beam

For an axial-symmetric lenslike system, the complex electric field on the output plane is related to the one on the input plane by an $ABCD$ ray matrix [2, 3]. Let the laser electric field on the input plane (the exit plane of a hollow-core fiber, for example) be given by[1] $E(\rho) = E_0\, J_0(2.405\rho/a)$ with $\rho \le a$, where ρ is the radial coordinate, E_0 the on-axis peak electric field, a the capillary radius, and J_0 the zero-order Bessel function of the first kind. The transverse electric field on the output plane, according to the diffraction theory in the paraxial approximation, is

[1] In principle $E(\rho)$ here can be replaced by a complex field as discussed in Sect. 2.4, the phase convention of the electric field should be taken with caution.

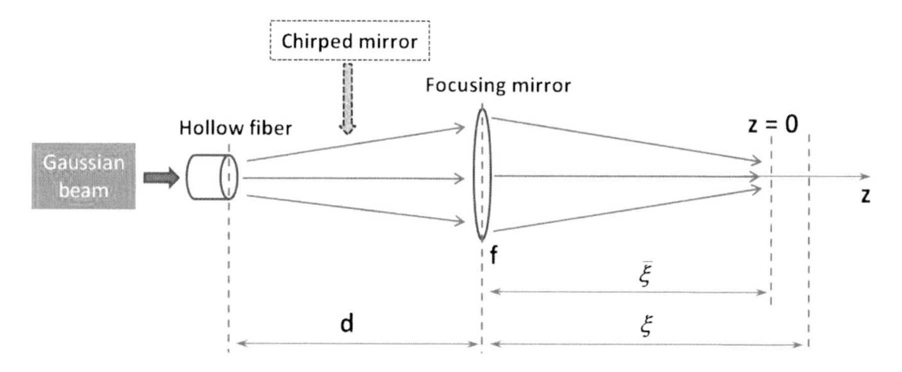

Fig. D.2 Sketch of the experimental setup for Type-1 Bessel beam generation [4]. Adapted from [6]. © (2012) by the American Physical Society

$$E_{\mathrm{TB}}(\xi, r) = E_0 \frac{-ik}{B(\xi)} \exp\left[ik(L + \xi + \frac{Dr^2}{2B(\xi)})\right] \times \int_0^a J_0\left(2.405\frac{\rho}{a}\right) J_0\left[\frac{kr\rho}{B(\xi)}\right] \exp\left[\frac{ikA(\xi)}{2B(\xi)}\rho^2\right] \rho d\rho, \tag{D.6}$$

where $k = 2\pi/\lambda_0$, and λ_0 is the central laser wavelength. The meanings of parameters in the equation will be defined explicitly below. Note that the integral in Eq. (D.6) becomes indeterminate if $B(\xi = \bar{\xi}) = 0$, where $\xi = \bar{\xi}$ is also the location of focus plane. As discussed in [3], the electric field at $\bar{\xi}$ can be written as

$$E_{\mathrm{TB}}(\bar{\xi}, r) = \frac{E_0}{A} \exp\left[ik(L + \bar{\xi} + \frac{Cr^2}{2A})\right] J_0(2.405\frac{r}{aA}). \tag{D.7}$$

For a lossless system, AD-BC=1. In the following I will show two truncated Bessel (TB) beams from different optical systems,[2] which have been used by Nisoli et al. [4] and Wörner et al. [5], respectively.

D.2.1 Tightly Focused Beam: Type-1 Bessel Beam

In the experiment of Nisoli et al. [4], the optical system setup is depicted in Fig. D.2. The radius of capillary is $a = 0.25$ mm, and the focal length of focus mirror is $f = 250$ mm. ξ and focus plane $\bar{\xi}$ are sketched in the figure. Laser pulse emerging from the hollow-core fiber propagates in free space for a distance $d = 2000$ mm [or L in Eq. (D.6)] to the focusing mirror where it further propagates for a distance ξ after the mirror to the output plane. The laser pulse is also compressed by the chirped

[2] The truncated Gaussian beam or other truncated beams can be constructed by modifying $E(\rho)$ on the input plane.

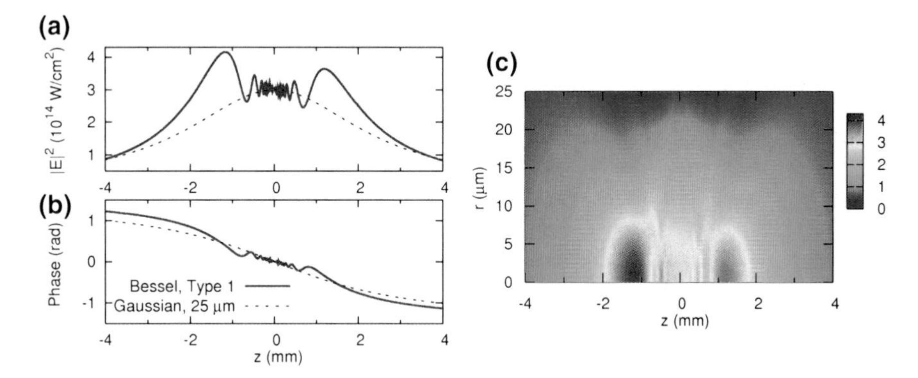

Fig. D.3 On-axis laser intensity **a** and phase **b** as a function of the propagation distance z: Type-1 Bessel (*solid lines*) versus Gaussian ($w_0 = 25\,\mu$m, *dashed lines*). **c** Spatial intensity distribution of Type-1 Bessel beam. Laser intensity at the focus is 3×10^{14} W/cm^2. Adapted from [6]. © (2012) by the American Physical Society

mirrors, but they are not included in the $ABCD$ matrix. For this optical system, the $ABCD$ matrix can be written as:

$$\begin{aligned}
A(\xi) &= 1 - \xi/f, \\
B(\xi) &= d + \xi(1 - d/f), \\
C &= -1/f, \\
D &= 1 - d/f.
\end{aligned} \tag{D.8}$$

The TB beam constructed by Eq. (D.8) is called Type-1 Bessel beam in this thesis. I plot the intensity $|E_{\mathrm{TB}}|^2$ and the phase ϕ_{TB} as a function of z for $r = 0$ (on-axis) in Fig. D.3a and b. Here the coordinate ξ has been replaced by $z = \xi$-$\bar{\xi}$ for convenience, and the phase ϕ_{TB} is set as 0 at $z = 0$ and $r = 0$ (focusing point). In the present case, $\bar{\xi} > f$, where $B(\bar{\xi}) = 0$ with B defined in Eq. (D.8). Laser wavelength $\lambda_0 = 780$ nm, and laser intensity at the focus is 3×10^{14} W/cm^2. For comparison, I fix the laser intensity at the focus and plot the intensity and the phase of a Gaussian beam with the beam waist $w_0 = 25\,\mu$m in Fig. D.3a and b, respectively. In Fig. D.3c, I plot the spatial distribution of laser intensity for the TB beam.

D.2.2 Loosely Focused Beam: Type-2 Bessel Beam

In the experiment of Wörner et al. [5], the setup is depicted in Fig. D.4. The hollow-core fiber (HCF) is similar to Nisoli et al.'s [4]. The beam coming out of the HCF (with radius $a = 0.125$ mm) is divergent. It is recollimated by a spherical mirror (focal length $f_1 = 1000$ mm) placed 1 m after the output of the HCF ($d_1 = 1000$ mm).

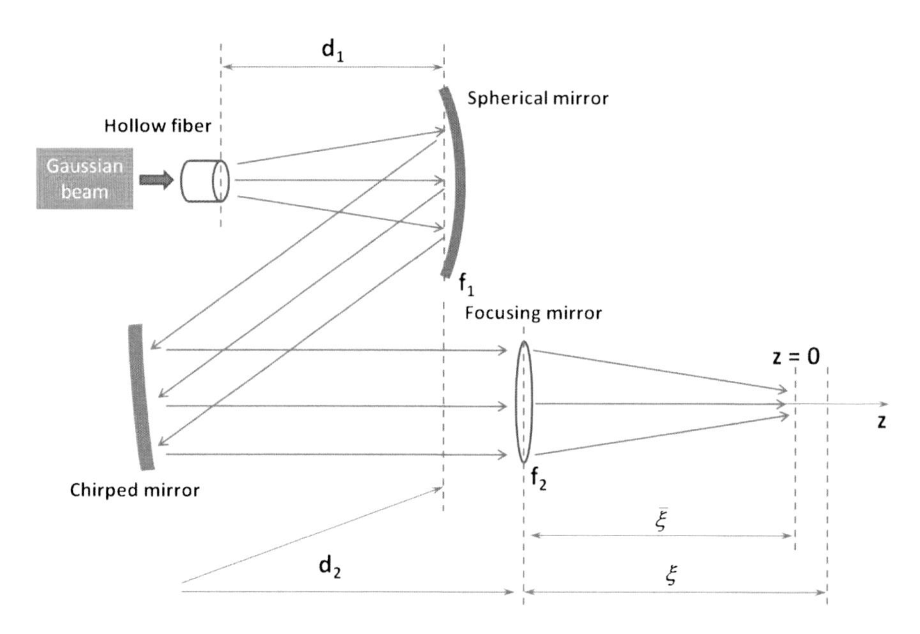

Fig. D.4 Sketch of the experimental setup for Type-2 Bessel beam generation. Adapted from [6]. © (2012) by the American Physical Society

The beam is then reflected 8 times on chirped mirrors and propagated a distance of 2 m from the spherical mirror ($d_2 = 2000\,\text{mm}$) until a focusing mirror (focal length $f_2 = 500\,\text{mm}$). It further propagates through a distance ξ after the mirror to the output plane. L in Eq. (D.6) equals to $d_1 + d_2$. I then write down the $ABCD$ matrix for this optical system without considering the chirped mirrors,

$$
\begin{aligned}
A(\xi) &= (1 - \frac{d_2}{f_1})(1 - \frac{\xi}{f_2}) - \frac{\xi}{f_1}, \\
B(\xi) &= (d_1 + d_2 - \frac{d_1 d_2}{f_1})(1 - \frac{\xi}{f_2}) - \xi(\frac{d_1}{f_1} - 1), \\
C &= -\frac{1}{f_1} - \frac{1}{f_2} + \frac{d_2}{f_1 f_2}, \\
D &= -\frac{d_1}{f_2} + (1 - \frac{d_1}{f_1})(1 + \frac{d_2}{f_2}).
\end{aligned}
\tag{D.9}
$$

To have a collimated laser beam before the focusing mirror f_2, it requires $d_1 = f_1$, i.e., the output of the HCF is put at the focal plane of spherical mirror f_1. In this case $\bar{\xi} = f_2$. The TB beam constructed by Eq. (D.9) is called Type-2 Bessel beam in this thesis. Similar to Fig. D.3, I plot on-axis intensity $|E_{\text{TB}}|^2$ and phase ϕ_{TB} as a function of z and the spatial distribution of intensity in Fig. D.5. In Fig. D.5a and b, I also plot

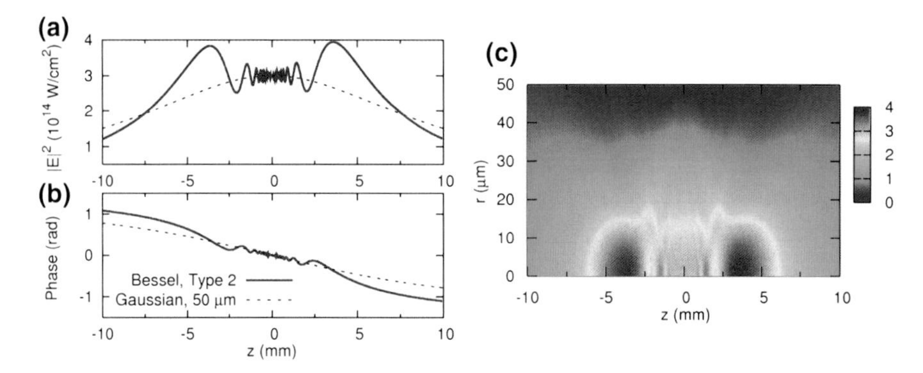

Fig. D.5 Same as Fig. D.3 except for loosely focused Type-2 Bessel and Gaussian ($w_0 = 50\,\mu$m) beams. Adapted from [6]. © (2012) by the American Physical Society

on-axis intensity and phase of a Gaussian beam with the beam waist $w_0 = 50\,\mu$m. The same laser wavelength and intensity (at the focus) are applied.

D.3 Ultrashort Laser Pulse and Geometric Phase

The spatial dependence of a laser field has been discussed above, I will write down a laser field with the complete spatial and temporal dependence in this section. In the moving coordinate frame (i.e., $z' = z$ and $t' = t - z/c$), the propagator term of e^{-ikz} can be eliminated, and the electric field can be written as

$$E_1(r, z', t') = Re\left[E_{\text{gau}}(r, z')A(r, z', t')e^{-i(\omega_0 t' + \varphi_{CE})}\right], \tag{D.10}$$

where

$$A(r, z', t') = \cos^2\{\frac{\pi[t' - \phi_{laser}(r, z')/\omega_0]}{\tau_p}\}. \tag{D.11}$$

Here the spatial beam is assumed as Gaussian, i.e., E_{gau}, it can be replaced by truncated Bessel or other beams straightforwardly. Carrier-envelope phase is represented by φ_{CE}, and τ_p in the envelope function $A(r, z', t')$ is the total duration of laser pulse, which equals to 2.75 times τ_w, the full width at half maximum (FWHM) of laser's intensity. If one could apply the Gaussian envelope in time domain, and then

$$A(r, z', t') = \exp\left[-(2\ln 2)\frac{(t' - \phi_{laser}(r, z')/\omega_0)^2}{\tau_w^2}\right]. \tag{D.12}$$

I also introduce the pulse energy W_{pulse} for laser beam:

$$W_{pulse} = \int\int I(r, z', t') 2\pi r dr dt', \tag{D.13}$$

where $I(r, z', t')$ is the spatial- and temporal-dependent laser intensity (assuming the cylindrical symmetry). For a Gaussian beam, there is an explicit expression of $I(r, z', t') = |E_1(r, z', t')|^2$ in Eq. (D.10), and one can derive an analytical expression for the pulse energy assuming the Gaussian envelope of Eq. (D.12) in time[3]:

$$W_{\text{pulse}} = I_0 \frac{\pi w_0^2}{2} \tau_w \sqrt{\frac{\pi}{4 \ln 2}}, \tag{D.14}$$

where I_0 is the laser peak intensity at the focus. If one chooses τ_p to be 3 cycles (7.8 fs), I_0=3×10^{14} W/cm^2, then W_{pulse} calculated from Eq. (D.14) is 24.45 μJ for the Gaussian beam in Fig. D.3. While W_{pulse} is 27.24 μJ for Type-1 Bessel beam in Fig. D.3 calculated numerically using Eq. (D.13).

If the fundamental laser field can be considered as propagating in the vacuum, its electric field can be expressed as an analytical form approximately by using Eq. (D.10). And then the propagation of high-harmonic field in the gas medium can be simplified. Let

$$t'' = t' - \varphi_{laser}(r, z')/\omega_0, \tag{D.15}$$

then

$$E_1(r, z', t'') = |E_{\text{gau}}(r, z')| \cos^2(\frac{\pi t''}{\tau_p}) \cos(\omega_0 t'' + \varphi_{CE}), \tag{D.16}$$

where the cosine-squared envelope is assumed, and it can be changed as the Gaussian envelope easily.

In order to solve Eqs. (2.55) and (2.62), the nonlinear polarization in the moving coordinate frame needs to be calculated. First one can compute $P_{nl}(r, z', t'')$ since in the time frame t'' the spatial component and the temporal part are separated. In other words, the fundamental laser field only depends on the peak field $|E_{\text{gau}}(r, z')|$. Using the Fourier transformation, one then could obtain

$$\widetilde{P}_{nl}(r, z', \omega) = \widehat{F}[P_{nl}(r, z', t')] = \widehat{F}[P_{nl}(r, z', t'')] e^{-i(\frac{\omega}{\omega_0})\phi_{laser}(r, z')}. \tag{D.17}$$

It can be seen in Eq. (D.17) that there are two contributions to the phase of nonlinear polarization: the first one is the induce-dipole phase, which depends only on laser peak intensity; the second is the geometric phase multiplied by the harmonic order. It is known that the most time-consuming job is the calculation of spatial dependent nonlinear polarization for atoms inside the medium as the harmonic field is propagated. It is the separation of the induced-dipole phase and the geometric phase that allows one to simplify the calculation. Using a batch of laser peak intensities, nonlinear polarizations in the time frame t'' are calculated and then stored. When it comes

[3] It is convenient to calculate the pulse energy at the focal plane where $z = 0$.

to solve the propagation equations for each value of ω, the nonlinear polarization in t'' for atoms or molecules inside the medium are obtained by the interpolation. Meanwhile the geometric phase is added up in order to transform the nonlinear polarization to the moving coordinate frame. The use of the interpolation method greatly improves the efficiency of harmonic field propagation. This approach is valid only for the multi-cycle laser pulse (FWHM, 10 optical cycles) applied.

References

1. R.W. Boyd, *Nonlinear Optics*, vol. 286, 2nd edn. (Academic Press, San Diego, 2003)
2. S.A.J. Collins, Lens-system diffraction integral written in terms of matrix optics. J. Opt. Soc. Am. A **60**, 1168–1177 (1970)
3. A. Yariv, Imaging of coherent fields through lenslike systems. Opt. Lett. **19**, 1607–1608 (1994)
4. M. Nisoli, E. Priori, G. Sansone, S. Stagira, G. Cerullo, S. De Silvestri, C. Altucci, R. Bruzzese, C. de Lisio, P. Villoresi, L. Poletto, M. Pascolini, G. Tondello, High-brightness high-order harmonic generation by truncated bessel beams in the sub-10-fs regime. Phys. Rev. Lett. **88**, 033902 (2002)
5. H.J. Wörner, H. Niikura, J.B. Bertrand, P.B. Corkum, D.M. Villeneuve, Observation of electronic structure minima in high-harmonic generation. Phys. Rev. Lett. **102**, 103901 (2009)
6. C. Jin, C.D. Lin, Comparison of high-order harmonic generation of Ar using truncated bessel and gaussian beams. Phys. Rev. A **85**, 033423 (2012)

Curriculum Vitae

Cheng Jin received the B.S. and M.S. degrees in Physics from Northwest Normal University, Lanzhou, Gansu, China, in 2003 and 2006, respectively and the Ph.D. degree in Physics from Kansas State University, Manhattan, Kansas, USA, in 2012.

Dr. Jin is currently a Research Associate in the Department of Physics at Kansas State University. His research is mainly about the strong-field physics, attosecond science, and ultrafast optics from the theoretical aspects, specifically including the development of macroscopic propagation model for high harmonics, the waveform synthesis for improving harmonic efficiency, the generation of an isolated attosecond pulse, the harmonic generation in a hollow-core waveguide, the photoionization of aligned molecules with high-harmonic light, and the molecular tunneling ionization.

C. Jin, *Theory of Nonlinear Propagation of High Harmonics Generated in a Gaseous Medium*, Springer Theses, DOI: 10.1007/978-3-319-01625-2,

Printed by Publishers' Graphics LLC
CAMZ130913.15.14.3